SYMBIOSIS IN THE SEA

THE BELLE W. BARUCH LIBRARY IN MARINE SCIENCE NUMBER 2

Symbiosis in the Sea

Edited by WINONA B. VERNBERG

Published for the Belle W. Baruch Coastal Research Institute by the

UNIVERSITY OF SOUTH CAROLINA PRESS

COLUMBIA, SOUTH CAROLINA

First Edition

Published in Columbia, South Carolina, by the

University of South Carolina Press, 1974

Manufactured in the United States of America by

Kingsport Press, Inc.

Library of Congress Cataloging in Publication Data

Main entry under title:

Symbiosis in the sea.

   (The Belle W. Baruch library in marine science,
no. 2)
   "This volume represents the second in a continuing
series of symposia on marine science, sponsored by the
Belle W. Baruch Coastal Research Institute."
   1. Symbiosis--Congresses. 2. Marine ecology--
Congresses. I. Vernberg, Winona B., 1924-   ed.
II. Belle W. Baruch Coastal Research Institute.
III. Title. IV. Series. [DNLM: 1. Marine biology--

Congresses. 2. Symbiosis--Congresses. QH548 S986
1974]

QH548.S95   574.5'24   73-16448
ISBN 0-87249-282-6

    This volume is dedicated to Dr. Raymond M. Cable in recognition of his contribution to science and for his boundless enthusiasm. He was born in Campton, Kentucky and received his undergraduate degree from Berea College; from there went to New York University for graduate training under the guidance of Dr. Horace W. Stunkard. After obtaining his Ph. D. in 1929, he returned to his undergraduate college for two years before joining the staff at Purdue University where he has remained since.

    Dr. Cable has been particularly interested in the evolutionary biology of the digenetic trematodes, and these studies have taken him to many different parts of the world. He has served as Vice-President and President of the American Society of Parasitologists, and was the recipient of a Guggenheim Fellowship. Berea College honored him with an honorary Sc. D. in 1955.

CONTRIBUTORS

B. W. ACHE, *Department of Biological Sciences, Florida Atlantic University, Boca Raton, Florida.*

R. M. CABLE, *Department of Biological Sciences, Purdue University, Lafayette, Indiana.*

T. C. CHENG, ANN CALI, AND DAVID A. FOLEY, *Institute for Pathobiology and Department of Biology, Lehigh University, Bethlehem, Pennsylvania.*

R. V. DIMOCK, JR., *Department of Biology, Wake Forest University, Winston-Salem, North Carolina.*

P. J. DECOURSEY, *Belle W. Baruch Coastal Research Institute, University of South Carolina, Columbia, South Carolina.*

R. W. GREENE, *Department of Biology, University of Notre Dame, Notre Dame, Indiana.*

J. B. JENNINGS, *Department of Pure and Applied Zoology, University of Leeds, Leeds, England.*

L. MUSCATINE, *Department of Zoology, University of California, Los Angeles, California.*

W. K. PATTON, *Department of Zoology, Ohio Wesleyan University, Delaware, Ohio.*

D. M. ROSS, *Faculty of Science, Office of the Dean, The University of Alberta, Edmonton, Canada.*

J. E. SIMMONS, *Department of Zoology, University of California, Berkeley, California.*

D. L. TAYLOR, *Rosenstiel School of Marine and Atmospheric Science, University of Miami, Miami, Florida.*

J. H. VANDERMEULEN, *Department of Zoology, Duke University, Durham, North Carolina.*

F. J. VERNBERG, *Belle W. Baruch Coastal Research Institute, University of South Carolina, Columbia, South Carolina.*

W. B. VERNBERG, *Belle W. Baruch Coastal Research Institute, University of South Carolina, Columbia, South Carolina.*

Symposium participants

F. J. Vernberg and R. M. Cable

B. W. Ache and D. M. Ross

B. James, D. L. Taylor, R. M. Cable
and P. J. DeCoursey

W. B. Vernberg
 and T. C. Cheng

J. E. Simmons
and L. Muscatine

W. K. Patton and R. V. Dimock

N. Mercando and J. B. Jennings

L. Muscatine
and R. W. Greene

PREFACE

The number of associations between different species of organisms living together is almost limitless. Such symbiotic relationships vary from temporary contact to metabolic dependence of one species upon another. The host represents the environment of the symbiont, and the symbionts have evolved adaptations to these environments in the same manner that free-living forms have adapted to their physical and biotic environment. Although many of the problems that symbionts must solve are basically the same as those of free-living animals, there are additional ones that the unencumbered cousins do not encounter, such as the resistance of the host species. Within the past few years more and more attention has been directed toward an understanding of the complexity of these relationships, and some of the most exciting work has been with marine organisms where such relationships are particularly common.

This volume represents the second in a continuing series of symposia on marine science, sponsored by the Belle W. Baruch Coastal Research Institute. The papers in Symbiosis in the Sea, the second in the series, covered a wide spectrum of symbiotic relationships and were directed toward one of two general areas of interest: 1) interactions between host and symbiont; and 2) evolutionary trends. The purpose of the symposium, Symbiosis in the Sea, was to highlight these recent advances with a view toward summarizing our present state of knowledge and identifying new research.

Special thanks are due to Drs. P. J. DeCoursey and F. J. Vernberg for editorial assistance, to Ms. Bettye Dudley for indexing, to Ms. Dorothy Knight for typing the manuscript and Ms. Margaret McLean for designing the jacket. The staff of the Belle W. Baruch Coastal Research Institute assisted in endless ways both in the organization of the symposium and in the preparation of the manuscript. Photographs are the courtesy of Dr. John M. Dean.

CONTENTS

Contents *(Cont.)*

SYMBIOSIS IN THE SEA

Influence of symbiotic algae on calcification in reef corals: critique and progress report

John H. Vandermeulen
Department of Zoology
Duke University
Durham, North Carolina

And

Leonard Muscatine
Department of Zoology
University of California
Los Angeles, California

Reef-building corals are invariably inhabited by symbiotic dinoflagellates (=zooxanthellae) (Yonge 1963). Experimental investigations show that these algae influence many aspects of reef coral biology, especially productivity (Roffman 1968), nutrient recycling (Pomeroy and Kuenzler 1969), and the rate of deposition of coral skeleton (aragonitic calcium carbonate)(Goreau 1959). The primary mechanism by which the algae exert their influence continues to be evaluated. This paper will describe some recent attempts to further understand the mechanism by which the zooxanthellae enhance reef-coral calcification.

The observations of Yonge (1931, 1940), Yonge and Nicholls (1931), Kawaguti and Sakumoto (1948), followed by the pioneering experiments of Goreau (1959) and Goreau and Goreau (1959), have demonstrated conclusively that light accelerates calcium deposition by reef corals, often as much as nine- or ten-fold. Recent experiments (Pearse and Muscatine 1971; Vandermeulen 1972) confirm these observations. This accelerating effect alone probably accounts for the spectacular success of reef-building by corals in shallow, well-illuminated tropical waters. There is strong evidence that the light-enhancing effect is due to photosynthesis by the symbiotic algae (Vandermeulen, Davis, and Muscatine 1972). There are at least three hypotheses which have been put forward to explain how algal photosynthesis might enhance calcification.

GOREAU'S HYPOTHESIS

Goreau (1959) observed that corals with algae in the light calcify at much higher rates than do those with algae in the dark, or those which have been caused to expel their zooxanthellae (Table 1). In addition, he noted that coral incubated in Diamox (2-acetyl-amino-1,3,4-thiadiazole-5-sulfonamide), an inhibitor of

TABLE 1
Calcium uptake in µg mg N^{-1} hr^{-1} by colonies of *Oculina diffusa* and *Manicina areolata* in presence and absence of zooxanthellae

Species	Light		Dark	
	With zooxanthellae	Without zooxanthellae	With zooxanthellae	Without zooxanthellae
O. diffusa	1.63± 3.38(7)*	0.37±0.01(6)	0.81± 0.15(9)	0.26±0.01(5)
M. areolata	462.00±63.20(11)	28.40±7.80(9)	71.70±14.90(8)	30.20±6.29(10)

*Number of samples in brackets (from Goreau, 1959).

carbonic anhydrase, and shown to be involved in mollusk shell formation (Freeman and Wilbur 1948; Wilbur and Jodrey 1955), calcify at
greatly reduced rates, both in the light or in darkness. Goreau
viewed an organic substrate between the calicoblastic epithelium and
the skeleton as the site of calcium carbonate precipitation. Calcium
taken up from seawater and transported to this site combines with
bicarbonate from respiration or from the environment. Goreau proposed the reaction scheme shown in Figure 1, stating that by removing
end-product carbon dioxide from the calcifying milieu by photosynthesis (equation 4), the zooxanthellae shift the reaction in favor of
increased precipitation of calcium carbonate (equation 2).

This scheme emphasizes the roles of calcium and CO_2, but neg-
lects the skeletal matrix as a constant and possibly important component of coral skeleton (Wainwright 1963; Young 1971).

$$(1) \quad Ca^{++} + 2HCO_3^- \longleftrightarrow Ca(HCO_3)_2$$

$$(2) \quad Ca(HCO_3)_2 \longleftrightarrow CaCO_3 + H_2CO_3$$

$$(3) \quad H_2CO_3 \xrightarrow{\text{carbonic anhydrase}} H^+ \downarrow + HCO_3^-$$

$$(4) \quad H_2CO_3 \xrightarrow{\text{carbonic anhydrase}} CO_2 + H_2O$$

Figure 1. Scheme of reactions involved in light-enhanced
calcification in corals. (After Goreau 1959)

PHOSPHATE REMOVAL HYPOTHESIS (CRYSTAL POISON THEORY)

Crystal growth of various calcium salts is inhibited by phosphate, presumably by disruption of normal crystal formation (for
full statement see Simkiss 1964c). Simkiss (1964a, b) pointed out
that the slightly higher values obtained by Goreau for coral calcification in the dark with zooxanthellae (Table 1) might be due to
removal of phosphate by the algae from the surface of the calcifying skeleton, suggesting that in the reef-building corals, light-
enhanced calcification is due to light-enhanced phosphate uptake by
the zooxanthellae. Marshalling of phosphate by zooxanthellae is
indicated by the observation that corals release less phosphate to
the environment than do marine animals of comparable size but lacking the algae (Pomeroy and Kuenzler 1969). Further, corals are
able to remove phosphate from solution when it is applied in relatively high amounts (Yonge and Nicholls 1931), and to a greater
extent in light than in darkness (Yamazato 1966).

"MATRIX" HYPOTHESIS

Organic matrices have been identified as a constituent of a
wide range of vertebrate and invertebrate mineralized tissues. Such
matrices are thought to function in nucleation and in the control
of skeletal form (Moss 1964). Wainwright (1963) first suggested
that, if the formation of a matrix is viewed as a limiting factor

in mineralization, then reduced organic carbon translocated from algae to host might serve directly in the synthesis of matrix or indirectly as an energy source for calcification or matrix synthesis, and thus account for light-enhanced mineralization by the coral.

Another aspect suggested by recent studies is the possible role of lipid in calcium binding. For example, lipids have been detected in association with skeletal organic constituents in gorgonian corals (Fox et al. 1969), in stony corals (Young, O'Connor, and Muscatine 1971), crustacean gastroliths (Travis 1960), and with calcium binding in bacteria (Ennever, Vogel, and Takazoe 1968). Further work in this area should prove fruitful.

A fourth, independent hypothesis which may be applied to reef coral calcification is that of Campbell and Speeg (1968) (see also Speeg and Campbell 1968). Based on studies of mineralization of shells of terrestrial gastropods, these investigators proposed the reaction:

$$NH_3 + HCO_3^- + Ca^{++} \longrightarrow CaCO_3 + NH_4^+$$

explaining that excretory ammonia combines with a proton released when bicarbonate is converted to carbonate. If applied to corals, one would have to explain how this reaction is accelerated by light. Increased removal of ammonium ions in the light, analogous to phosphate removal, is one possibility.

While the foregoing hypotheses are not necessarily mutually exclusive, none has actually been put to a critical test, and therefore all remain open to question. For example, in Goreau's scheme it is difficult to envisage how carbon dioxide can be removed from the calcifying milieu by photosynthesis without also removing the carbon dioxide going into carbonate formation (Fig. 1, reaction 1). Goreau interpreted the effect of Diamox as inhibiting carbonic anhydrase activity in coral animal tissues. However, some algae depend on carbonic anhydrase for utilization of bicarbonate in photosynthesis (Hatch, Osmond, and Slatyer 1971). If this is true for zooxanthellae, then in Goreau's experiments Diamox may actually have inhibited photosynthesis in addition to its inhibition of carbonic anhydrase activity. Further, if carbon dioxide removal results in simple physico-chemical precipitation of calcium carbonate, then one should expect calcareous marine algae to calcify uniformly over their entire surface (Lewin 1962). They do not. In fact, only certain parts of the surface of some calcareous algae calcify while the remaining surface does not (Lewin, 1962; Stark, Almodovar, and Kraus 1969; Pearse 1972). Goreau (1963) also observed that the tips of the branching coral *Acropora*, which are virtually algae-free, are the fastest growing portions of the skeleton. These tips also have the highest light:dark ratio of calcification (Pearse and Muscatine 1971). Finally, Vandermeulen (1972) has studied the location and distribution of zooxanthellae in a variety of reef corals. His data (Table 2, Fig. 2) for three common reef-building corals show that in the coenosteal tissues (which overlie a significant amount of the coral surface, and are involved in continuous skeletal growth) the

TABLE 2
Distribution of zooxanthellae in gastrodermis of three reef-building corals. Counts made in alternating 10 micron sections (N) prepared from randomly sampled coral colonies. Zooxanthellae/125 microns (from Vandermeulen, 1972)

| Species | Sample | N | Coenosteum | | Polyp |
			Superior gastrodermis	Inferior gastrodermis	Basal gastrodermis
Pocillopora damicornis	1	25	15.44 ± 4.17[*]	0.68 ± 0.26	4.36 ± 1.90
	2	15	11.26 ± 1.99	1.15 ± 1.09	4.37 ± 1.94
	3	25	11.24 ± 2.07	0.03 ± 0.02	3.72 ± 1.61
Seriatopora sp.1		17	20.29 ± 2.40	1.94 ± 1.53	7.58 ± 5.15
	2	15	15.33 ± 4.32	2.06 ± 1.73	6.80 ± 4.89
Acropora sp.	1	15	20.00 ± 8.75	1.93 ± 1.73	4.60 ± 3.48
	2	15	9.86 ± 3.75	0.01 ± 0.34	6.40 ± 3.50

[*]mean ± standard deviation.

majority of the algae are located in the oral gastrodermis, and that very few are found in the aboral gastrodermis adjacent to the calicoblast epidermis. This means that the bulk of the zooxanthellae are about 100μ farther from the site of calcification then previously supposed. Enhancement of calcification by removal of carbon dioxide, phosphate ions, or ammonium ions from the calcifying milieu would require maintenance of a strong gradient of these substances from point of calcification to location of the algae. This gradient must traverse the coelenteron and its dynamic fluid interior. An axial gradient of several millimeters would be necessary in the tips of *Acropora*. These arguments seem contrary to the idea that light-enhanced calcification can be accounted for by simple physiocochemical precipitation.

TESTING THE GOREAU HYPOTHESIS

We attempted to test Goreau's hypothesis by inhibiting specifically the removal of CO_2 by the algae during calcification, without affecting oxygen production or photophosphorylation. The results are published elsewhere (Vandermeulen, Davis, and Muscatine 1972) and are only briefly summarized here. Our rationale was based on the scheme for electron transport in photosynthesis and the inhibitors used in connection with this scheme (see Izawa et al 1967). We attempted to use methyl viologen (MV) to inhibit CO_2 reduction, but soon found that MV in high enough concentration to inhibit photo-

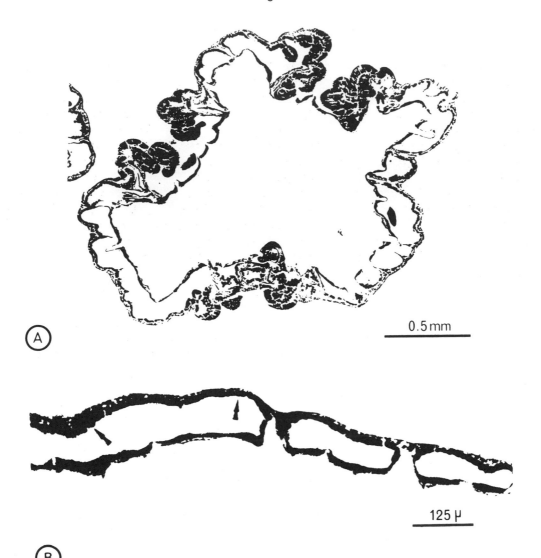

Figure 2. Light micrograph of section through decalcified and paraffin-embedded adult *P. damicornis*, showing arrangement of polyps with interconnecting coenosteum (A) and distribution of zooxanthellae (arrows) in the coenosteum.

synthesis killed corals outright, and that at low concentrations *in vivo* it had no effect at all. Studies with Diuron (DCMU) and Monuron (CMU) were more successful: at concentrations of 10^{-4} to 10^{-5} these compounds inhibited algal photosynthesis completely, and coral calcification was correspondingly depressed to intrinsic dark levels. However, this approach succeeded only in confirming that photosynthesis and not some other photobiological effect mediates light-enhanced calcification.

TESTING THE "MATRIX" HYPOTHESIS

An important part of the test of this hypothesis is marshalling the evidence both for the presence of a discrete organic matrix in coral skeletons and for the translocation of organic material from algae to matrix. Wainwright (1963) observed a fibrillar complex by light and electron microscopy after *Pocillopora damicornis* skeleton was decalcified in ethylenediaminetetraacetic acid (EDTA). Using X-ray diffraction he identified chitin as a major constituent of this residue, which he referred to as the organic matrix. He obtained no evidence for the presence of protein. Using different methods with *Pocillopora* skeleton, Young (1969, 1971) obtained acid-insoluble material after decalcification. Hydrolysis of this material yielded glucosamine and a wide range of amino acids. A survey of corals from four suborders indicated that chitin was present mainly in the genus *Pocillopora*, while all samples gave evidence for the presence of proteinaceous constituents. Young's data strongly suggest that the organic material consistently obtained after decalcification of coral skeletons is a true skeletal matrix.

Further evidence for the existence of a matrix comes from ^{14}C-tracer studies. It is now well established that photosynthetically-fixed carbon moves from zooxanthellae to host in a variety of associations. The principal mobile compounds are glycerol, alanine, and glucose (Muscatine and Cernichiari 1969; Smith, Muscatine, and Lewis 1969; Trench 1971a, b, c; Lewis and Smith 1971). If corals are incubated in the light for 24 hrs with $^{14}CO_2$, 45% of the fixed ^{14}C is detected in the animal tissue fraction and about 10% is recoverable from the skeleton. A small fraction of the skeletal ^{14}C occurs in the matrix (Muscatine and Cernichiari 1969). In this fraction, a significant amount of ^{14}C occurs as amino acids, particularly serine-^{14}C and methionine-^{14}C, and a wax ester, cetyl palmitate-^{14}C (Young, O'Connor, and Muscatine 1971). The possibility that the observed matrix is an artifact seems highly unlikely, in view of the fact that a $^{14}CO_2$ tracer can label the matrix in relatively short incubations and to a greater extent in light than in darkness. Young (1969) attempted to inhibit matrix protein synthesis in *P. damicornis* with Puromycin but could not obtain clear evidence for a positive effect of the inhibitor. However, he did obtain data which suggested that ^{14}C incorporation into matrix was correlated strongly with ^{14}C incorporation into carbonate (r=0.63, p <0.01). Recent experiments on *Acropora* indicate that high rates of calcification in algae-free tips may be explained by axial translocation of organic material from algae located farther down the branch (Pearse and Muscatine 1971). These various observations, from X-ray diffraction, electron microscopy, biochemistry, and radioisotopic tracer studies, strongly indicate a true organic matrix in intimate association with the calcareous skeleton.

To test the hypothesis that in the light the zooxanthellae enhance calcification by augmenting matrix formation, coral branches were incubated in the dark in seawater enriched with glycerol, alanine, and glucose. The aim was to enhance calcification in the dark to levels normally associated with light-enhanced calcification. The feasibility of this approach was suggested by the observation that the corals readily absorb glycerol, alanine, and glucose from seawater (Lewis and Smith 1971; Stephens 1962).

MATERIALS AND METHODS

Experimental Procedures

Collection and maintenance of experimental animals
Heads of the colonial branching coral *P. damicornis* were collected from a reef in Kaneohe Bay, Oahu, Hawaii, in depths of 1 to 5 m. Only colonies of uniform coloration, suggesting even distribution of zooxanthellae, and of normal growth form were selected. Specimens were transferred under water to collecting buckets, and on return to the laboratory were maintained in shallow tanks of running seawater at the Hawaii Institute of Marine Biology on Coconut Island, Hawaii. Generally collections were made the same day of an experiment. Occasionally, corals were collected the evening before and kept overnight in the seawater system.

Sampling of coral branches
Sampling was done by breaking off branch tips from 5 to 15 mm in length with small side-cutting pliers. Branches of *P. damicornis* were always taken from the central region of the head to minimize possible variation in calcification rates or photosynthesis rates with respect to position of branches within the coral colony (C. Clausen, personal communication). The peripheral branches, which are often shaded and of nonuniform coloring were avoided.

Incubation conditions
Intact corals were incubated in 15 ml filtered seawater ("Aquapure" filters, 5μ pore size) in 30 ml Griffin beakers. The beakers were placed in a plexiglass chamber illuminated from below and the sides by fluorescent lights delivering ca. 20,000 lux at the bottom of the beakers (Weston Illumination Meter, No. 756, quartz filter). The beakers were covered on top with aluminum foil to aid in reflection of light onto the corals. In dark experiments, the beakers containing the experimental animals and incubation media were placed in small black plastic bags, wrapped in aluminum foil, and then placed in the plexiglass light chamber. Sampling of dark experimentals was done in dim light (less than 500 lux at the tissues). Temperature in the incubation chamber was maintained at 26°C $\pm$ 1°C (based on Clausen 1971) by constant water circulation.

Glycerol, alanine, and glucose enrichment
Individual stock solutions of glycerol, alanine, and glucose were made up in distilled water in concentrations isotonic with

seawater. Experimental solutions were prepared freshly before each experiment by addition of 1 ml stock solution to an appropriate volume of filtered seawater.

Incubation in ^{14}C and ^{45}Ca

^{14}C was added to the incubating medium as $NaH^{14}CO_3$ (New England Nuclear, 60 μCi/mMol) to give a final activity of 1 μCi/ml. Incubations in ^{14}C were for 30 min. In calcification studies, $^{45}CaCl_2$ (New England Nuclear) was added to give a final activity of 3 μCi/ml. Incubation times in ^{45}Ca were from 30 min to 4 hrs.

Samples were routinely preincubated in the appropriate experimental solutions for 1 hr before addition of radioisotope. Following incubations in isotope, the experimental organisms were washed in filtered seawater until washings contained negligible radioactivity.

Analytical Procedures

Tissue-^{14}C

Coral branches, following incubation and rinsing, were heated in concentrated NH_4OH to digest organic matter. NH_4OH was used since it could be evaporated from aliquots of digest and so not contribute to sample self-absorption. Aliquots of the resultant alkaline tissue digest were assayed for radioactivity and for protein nitrogen.

Preparation of skeletal-^{45}Ca

After calcification experiments, the exposed end of each branch was cut off with side-cutting pliers and discarded to avoid error due to uptake of ^{45}Ca by equilibrium exchange during the incubation. Coral tissue was removed with hot alkali and assayed for protein nitrogen. The remaining clean skeleton samples were dried to constant weight, weighed, and dissolved in 6N HCl for assay of ^{45}Ca.

Radioassay

For measurement of ^{14}C, 100 μl aliquots of tissue digest were dried on nickel-plated planchets (concentric) and treated with 0.1N HCl to remove unincorporated ^{14}C-bicarbonate. For ^{45}Ca assay, 100 μl samples of dissolved skeleton were plated onto planchets, dried, and kept at 60°C over a hotplate to avoid deliquescence.

Planchets were counted with an end-window GM tube (LND 733) and scaler (Nuclear Supplies, model SA-250). All counts were corrected for background, and in the case of skeletal-^{45}Ca, for sample self-absorption (cf. Chase and Rabinowitz 1968). Activity was expressed as counts x min^{-1} x μg protein $nitrogen^{-1}$ (cpm/μg.Pr.N.).

Protein-nitrogen determination

Protein nitrogen (Pr.N) was measured by the method of Lowry et al. (1951) using bovine serum albumin standards (Armour, protein standard solution, 10.1 mg (Pr.N per ml).

RESULTS

Demonstration of Light-enhanced Calcification

Coral branches were incubated in ^{45}Ca in the light and in the dark for up to 23 hrs, and skeletal incorporation of ^{45}Ca was measured periodically. The results (Fig. 3) show that after 24 hrs ^{45}Ca deposition in the light is about 7.5 times greater than in the dark. The results further show that although the light-dark differences in ^{45}Ca uptake are evident after 30 min of incubation (first two points in Fig. 3), this difference is not clearly established until after 2 hrs of incubation. To ensure light-dark differences in calcification experiments were carried out for 30 min to 4 hrs.

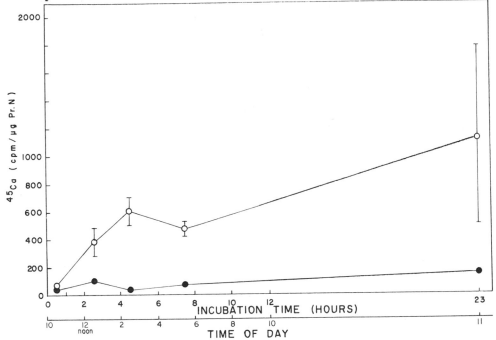

Figure 3. Calcification in branches of *P. damicornis* incubated continuously in ^{45}Ca in the light (open circles) or in the dark (closed circles). Means ± standard error for three replicates.

Effect of Incubation with Glycerol, Alanine, and Glucose on Coral Calcification

1% organic enrichment

Coral branches were incubated for 4 hrs in the dark in seawater enriched with 1% each of glycerol, alanine, and glucose. Controls consisted of branches incubated for the same period either in the light or in the dark, but without organic enrichment. The results, summarized in Figure 4, show that enhancement was not obtained in coral incubated in the dark in the presence of the combined specimens incubated in the dark in nonenriched seawater, i.e. intrinsic dark calcification.

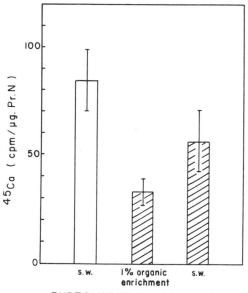

Figure 4. Effect of a 1% solution of glycerol, alanine, and glucose on calcification in the dark in branches from *P. damicornis*. Mean ± standard error for five replicates. Shading denotes dark incubations.

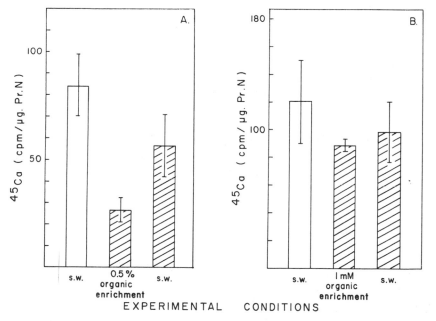

Figure 5. Effect of (A) 0.5% organic enrichment, and (B) 1 mM organic enrichment on coral calcification in the dark. Mean ± standard error for five replicates. Shading denotes dark incubations.

0.5% and 1 mM organic enrichment

To explore the possibility that a 1% organic enrichment may have possibly inhibited ^{45}Ca uptake due to a concentration effect, the experiment was repeated, but with decreased concentrations of glycerol, alanine, and glucose. The results (Fig. 5A and B) show that decreasing the concentration had some effect. Calcification at enrichment concentrations of 1 mM was similar to that obtained in dark controls in seawater only.

Enrichment without glycerol

Glycerol is known to have a disruptive effect on symbiotic algae *in situ* (Whitney 1907). Thus, *Hydra viridis*, a hydrozoan with symbiotic zoochlorellae, readily loses its total algal flora when incubated in 0.5% glycerol over a period of seven to 10 days (Muscatine 1961). To test for this alternative, coral branches were incubated in the dark in seawater labeled with ^{45}Ca, and enriched with 1 mM each of glucose and alanine. The results were compared with light and dark controls and are summarized in Figure 6. Under these conditions no enhancement was obtained above normal dark intrinsic levels.

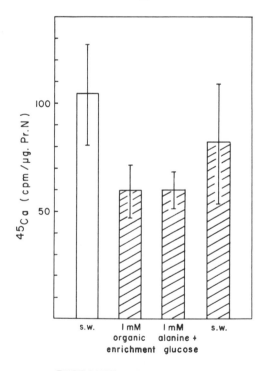

EXPERIMENTAL CONDITIONS

Figure 6.   Effect on coral calcification of 1 mM organic enrichment from which glycerol is omitted. Mean ± standard error for five replicates. Shading denotes dark incubations.

"Pulse-chase" in the dark to test for inhibition
of ^{45}Ca uptake by added organic substrates

The results obtained so far suggest that ^{45}Ca uptake is inhibited by the presence of added glucose, glycerol, and alanine in seawater. To circumvent this possibility, coral branches were preincubated in the dark for 2 hrs in seawater enriched with 1 mM each of glycerol, alanine, and glucose. Following this, the branches were transferred to nonenriched seawater in the dark and "chased" with ^{45}Ca for 4 hrs. The results of this experiment show no increase in ^{45}Ca uptake in the experimental samples under these conditions when compared with both light and dark controls (Fig. 7).

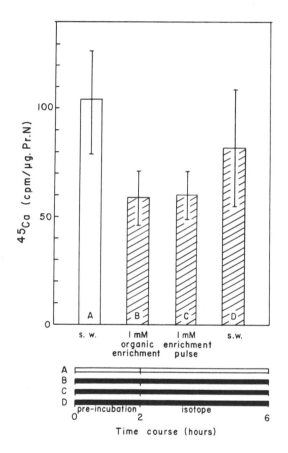

Figure 7. Effect of a 2-hr preincubation in 1 mM organic enrichment
(pulse) followed by transfer to nonenriched seawater containing ^{45}Ca (chase). Shading denotes dark incubations.
Mean ± standard error for five replicates.

The results thus far show that organic enrichment not only fails to enhance calcification but also inhibits intrinsic dark calcification to a certain extent. The data further indicate that

enhanced calcification may indeed have a light requirement. One possibility is that light absorption is necessary for isomerization of
matrix monomers and subsequent matrix organization, similar to structural conformational changes effected by light in the chromophore
moiety of visual pigments (Thomas 1965). This possibility was tested
by incubating coral branches in enriched seawater in the light in
the presence of 5 x 10^{-4} M DCMU, demonstrated inhibitory to algal
photosynthesis (see Vandermeulen, Davis, and Muscatine 1972). The
results (Fig. 8) show no increased ^{45}Ca uptake under these conditions above intrinsic dark levels of calcification.

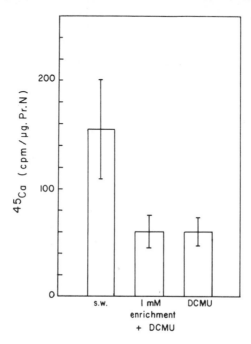

EXPERIMENTAL   CONDITIONS

Figure 8. Effect of 1 mM organic enrichment on coral calcification
in the light in the presence of 5 x 10^{-4} M DEMU. Mean ±
standard error for five replicates.

Incubation in enriched seawater in the light

To determine whether actively photosynthesizing algae are essential for enhanced calcification in addition to the added glycerol,
alanine, and glucose, coral branches were incubated in enriched seawater in the light. The results show that normal light-enhanced
calcification is depressed to approximately 40% of the seawater control level (Fig. 9).

When, under these same conditions of light plus-enriched seawater, photosynthesis also was monitored (by measuring ^{14}C fixation
in coral branches incubates in ^{14}CO$_2$). ^{14}C fixation was found to be
70% below normal light controls, intermediate between light controls
incubated in 5 x 10^{-4}M DCMU (Fig. 10). Therefore, the inhibitory
effect of organic enrichment in the light is on photosynthesis rather
than on calcification.

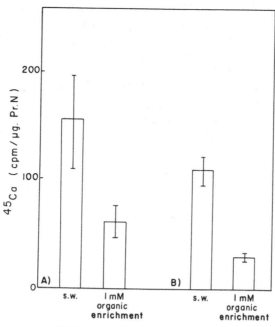

Figure 9. Effect of 1 mM organic enrichment on coral calcification in the light. Results shown are from two separate experiments performed on branches from separate colonies. Mean ± standard error for five replicates. 20,000 lux.

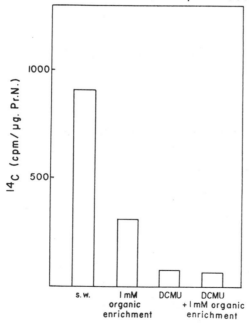

Figure 10. Effect of 1 mM organic enrichment of ^{14}C fixation. 1-hr incubation in organic enriched seawater, or enriched seawater plus 5×10^{-4} M DCMU. Means only of two replicates.

DISCUSSION

The results indicate that the organic enrichment experiments as designed here did not adequately test the "matrix hypothesis." We can offer no clear-cut explanation for the small suppression of calcification in the dark by organic enrichment. The observed depression of ^{14}C fixation in the light in the presence of organic enrichment might reflect an isotope dilution effect, with tracer ^{14}C being diluted by increased respiratory ^{12}C becoming available from respiration of excess glucose, alanine, and glycerol. Such a dilution effect has been described for *Chlorohydra viridissima* following feeding (Szmant 1971). However, this does not seem likely here since ^{14}C fixation more probably reflects a decrease in algal photosynthesis, resulting from the exogenously supplied organic compounds. In addition, by partially inhibiting photosynthesis, phosphate removal may also be partially inhibited. If the phosphate crystal poison theory (Simkiss 1964a, b) holds for coral calcification, the failure to remove these ions in the light would render equivocal any interpretation of the effect of exogenous organic substrates on calcification. Possibly a critical experiment, overlooked in these studies, is a pulse-chase experiment, including not only a pulse with organic enrichment in the dark, but a chase in a nonenriched seawater in the light. If light-enhanced calcification involves not only algal supply of glycerol, alanine, and glucose as substrates for animal matrix precursor production, but also phosphate removal from the animal tissues and the calcifying milieu, then one would expect vastly enhanced calcification during the chase phase of this experiment.

In summary, we have given a brief overview of the status of our knowledge of how symbiotic algae accelerate calcification in reef corals and an account of recent work in progress. In our view, the next decade of research in this area will very likely profit from more clearly defined conceptual models of coral calcification.

ACKNOWLEDGMENT

The studies described herein were supported by a grant from the National Science Foundation (NSF 11940), and an Edwin Pauley Fellowship. We gratefully acknowledge the assistance of the staff of the Hawaii Institute of Marine Biology. We thank Dr. Vicki Buchsbaum Pearse for critical review of the manuscript.

LITERATURE CITED

Campbell, J. W., and Speeg, K. V., Jr. 1968. Theoretical considerations of the possible role of ammonia in the biological deposition of calcium carbonate. Am. Zool 8:770.

Chase, G. D., and Rabinowitz, J. L. 1968. _Principles of radioiso-tope methodology_. 3rd ed. Minnesota: Burgess Publishing Co.

Clausen, C. 1971. Effects of temperature on the rate of ^{45}Ca up-take by _Pocillopora damicornis_. In _Experimental coelenterate biology_, ed. H. M. Lenhoff, L. Muscatine, and L. V. Davis, pp. 246-59. Honolulu: Univ. of Hawaii Press.

Ennever, J., Vogel, J., and Takazoe, I. 1968. Calcium binding by a lipid extract of _Bacterionema matruchotii_. _Calc. Tiss. Res._ 2:296-98.

Fox, D. L., Smith, V. E., Grigg, R. W., and MacLeod, W. D. 1969. Some structural and chemical studies of the microspicules in the fan-coral _Eugorgia ampla_ Verrill. _Comp. Biochem. Physiol._ 28:1103-14.

Freeman, J. A., and Wilbur, K. M. 1948. Carbonic anhydrase in molluscs. _Biol. Bull. mar. lab. Woods Hole_ 94:55-59.

Goreau, T. F. 1959. The physiology of skeleton formation in corals. I. A method for measuring the rate of calcium deposition by corals under different conditions. _Biol. Bull. mar. lab. Woods Hole_ 116:59-75.

_____. 1963. Calcium carbonate deposition by coralline algae and corals in relation to their roles as reef-builders. _Ann. N. Y. Acad. Sci._ 109:127-67.

_____, and Goreau, N. I. 1959. The physiology of skeleton forma-tion in corals. II. Calcium deposition by hermatypic corals under various conditions in the reef. _Biol. Bull. mar. lab. Woods Hole_ 117:239-50.

Hatch, M. D., Osmond, C. B., and Slatyer, R. O. 1971. _Photosynthe-sis and photorespiration_. New York: Wiley-Interscience.

Izawa, S., Connolly, T. N., Winget, G. D., and Good, N. E. 1967. Inhibition and uncoupling of photophosphorylation in chloro-plasts. In _Energy conversion by the photosynthetic apparatus_. Brookhaven Symposia in Biology, No. 19, pp. 169-187. U. S. Dept. of Commerce, Springfield, Va.

Kawaguti, S., and Sakumoto, D. 1948. The effect of light on the calcium deposition of corals. _Bull. oceanogr. Inst. Taiwan_ 4: 65-70.

Lewin, J. 1965. Calcification. In _Physiology and biochemistry of algae_, ed. R. Lewin, Ch. 28. New York: Academic Press.

Lewis, D. H., and Smith, D. C. 1971. The autotrophic nutrition of symbiotic marine coelenterates with special reference to hermatypic corals. I. Movement of photosynthetic products between the symbionts. _Proc. Roy. Soc._ (B) 178:111-29.

Lowry, O. H., Rosebrough, N. J., Farr, A. L., and Randall, R. 1951. Protein measurement with the Folin phenol reagent. _J. Biol. Chem._ 193:265-75.

Moss, M. 1964. The phylogeny of mineralized tissues. _Internat. Rev. Gen. Exper. Zool._ 1:297-331.

Muscatine, L. 1961. Symbiosis in marine and fresh water coelen-terates. In _The biology of hydra and some other coelenterates_, ed. H. M. Lenhoff and W. F. Loomis, pp. 255-68. Miami: University of Miami Press.

_____, and Cernichiari, E. 1969. Assimilation of photosynthetic products of zooxanthellae by a reef coral. Biol. Bull. mar. lab. Woods Hole 137:506-23.

Pearse, V. B. 1972. Radioisotopic study of calcification in the articulated coralline alga *Bossiella orbigniana*. J. Phycol. 8:88-97.

_____, and Muscatine, L. 1971. Role of symbiotic algae (zooxanthellae) in coral calcification. Biol. Bull. mar. lab. Woods Hole 141:350-63.

Pomeroy, L. R., and Kuenzler, E. J. 1969. Phosphorus turnover by coral reef animals. Proc. 2nd Conf. on Radioecology, AEC Conf-670503: 474-82.

Roffman, B. 1968. Patterns of oxygen exchange in some Pacific corals. Comp. Biochem. Physiol. 27:405-18.

Simkiss, K. 1964a. Possible effects of zooxanthellae on coral growth. Experientia 20:140.

_____. 1964b. The inhibitory effects of some metabolites on the precipitation of calcium carbonate from artificial and natural seawater. J. Cons. Int. Explor. Mer. 29:6-18.

_____. 1964c. Phosphates as crystal poisons of calcification. Biol. Rev. 39:487-505.

Smith, D. C., Muscatine, L., and Lewis, D. 1969. Carbohydrate movement from autotrophs to heterotrophs in parasitic and mutualistic symbiosis. Biol. Rev. 44:17-90.

Speeg, K. V., Jr., and Campbell, J. W. 1968. Formation and volatilization of ammonia gas by terrestrial snails. Am. J. Physiol. 214:1392-1402.

Stark, L., Almodovar, L., and Kraus, R. W. 1969. Factors affecting the rate of calcification in *Halimeda pountia* (L.), Lanouroux and *Halimeda discoidea* Decaiene. J. Phycology. 5: 305-12.

Stephens, G. C. 1962. Uptake of organic material by aquatic invertebrates. I. Uptake of glucose by the solitary coral *Fungia scutaria*. Biol. Bull. mar. lab. Woods Hole 123:648.

Szmant, A. M. 1971. Patterns of $^{14}CO_2$ uptake by *Chlorohydra viridissima*. In Experimental cell biology, ed. H. M. Lenhoff, L. Muscatine, and L. V. Davis, pp. 192-201. Honolulu: Univ. of Hawaii Press.

Thomas, J. B. 1965. Primary photoprocesses in biology. Amsterdam: North-Holland Publishing Co.

Travis, D. F. 1960. Matrix and mineral deposition in skeletal structures of the decapod crustacea (phylum Arthropoda). In Calcification in biological systems, ed. R. F. Sognnaes, pp. 57-116. Washington, D. C.: Amer. Assoc. Advancement Science.

Trench, R. K. 1971a. The physiology and biochemistry of zooxanthellae symbiotic with marine coelenterates. I. The assimilation of photosynthetic products of zooxanthellae by two marine coelenterates. Proc. Roy. Soc. (B) 177:225-35.

_____. 1971b. The physiology and biochemistry of zooxanthellae symbiotic with marine coelenterates. II. Liberation of fixed

^{14}C by zooxanthellae *in vitro*. Proc. Roy. Soc. (B) 177:237-50.

_____. 1971c. The physiology and biochemistry of zooxanthellae symbiotic with marine coelenterates. III. The effect of homogenates of host tissues on the excretion of photosynthetic products *in vitro* by zooxanthellae from two marine coelenterates. Proc. Roy Soc. (B) 177:251-64.

Vandermeulen, J. H. 1972. Studies on skeleton formation, tissue ultrastructure, and physiology of calcification in the reef coral *Pocillopora damicornis* Lamarch. Ph.D. dissertation, University of California.

_____, Davis, N., and Muscatine, L. 1972. The effect of inhibitors of photosynthesis on zooxanthellae in corals and other marine invertebrates. Mar. Biol. 16:185-91.

Wainwright, S. A. 1963. Skeletal organization in the coral, *Pocillopora damicornis*. Quart. J. Microsc. Sci. 104:169-83.

Whitney, D. D. 1907. Artificial removal of the green bodies of *Hydra viridis*. Biol. Bull. mar. lab. Woods Hole 13:291-99.

Wilbur, K. M., and Jodrey, L. 1955. Studies on shell formation. V. The inhibition of shell formation by carbonic anhydrase inhibition. Biol. Bull. mar. lab. Woods Hole 108:359-65.

Yamazato, K. 1966. Calcification in a solitary coral, *Fungia scutaria*. Ph.D. dissertation, Honolulu: Univ. of Hawaii.

Yonge, C. M. 1931. The significance of the relationship between corals and zooxanthellae. Nature 128:309-11.

_____. 1940. The biology of reef-building corals. Sci. Rep. Gt. Barrier Reef Exped. 1:353-91

_____. 1963. The biology of coral reefs. In Advances in Marine Biology, ed. F. S. Russell, Vol. I, pp. 209-60. New York: Academic Press.

_____, and Nicholls, A. R. 1931. Studies on the physiology of corals. V. The effect of starvation in light and darkness on the relationship between corals and zooxanthellae. Sci. Rep. Gt. Barrier Reef Exped. 1:179-211.

Young, S. D. 1969. Studies on the skeletal organic material in hermatypic corals with emphasis on *Pocillopora damicornis*. Ph.D. dissertation, University of California.

_____. 1971. Organic material from scleractinian coral skeletons. I. Variation in composition in between several species. Comp. Biochem. Physiol. 40B:113-20.

_____, O'Connor, J. D., and Muscatine, L. 1971. Organic material from scleractinian coral skeletons. II. Incorporation of ^{14}C into protein, chitin and lipid. Comp. Biochem. Physiol. 40B: 945-58.

Mollusc.
~ opisthobranchiata ~ sea slugs
eg Elysia viridis

Sacoglossans and their chloroplast endosymbionts

Richard W. Greene
Department of Biology
University of Notre Dame
Notre Dame, Indiana

Since the work of De Negri and De Negri in 1876, it has been recognized that members of the opisthobranch Order Sacoglossa carry chlorophyll-like pigments in their bodies. Historically, the presence of the pigments has been attributed to ingested food (Fretter 1941) or to symbiotic algae (Kawaguti 1941; Kawaguti, Yamamoto, and Kamishima 1965; Ostergaard 1955; Yonge and Nicholas 1940; Kay 1968). The first indication that another association might be involved appeared when Kawaguti and Yamasu (1965) identified the green bodies in Japanese *Elysia atroviridis* as chloroplasts of the green alga, *Codium fragile*.

This original report precipitated a number of reports of additional associations involving sacoglossan slugs with algal chloroplasts (Taylor 1967, 1968; Trench, Greene, and Bystrom 1969; Trench 1969; Greene 1970 a, b). Table 1 lists the species of slugs investigated which carry chloroplast symbionts, as well as their sources of chloroplasts if they have been identified.

ACQUISITION OF CHLOROPLAST SYMBIONTS

All members of the Order Sacoglossa feed suctorially (Fretter 1941) and, with but one exception, they feed on algae. The animals creep along on a particular algal substrate and pierce individual plant cells with a single radular tooth. Once an algal cell is

TABLE 1
Sacoglossans known to harbor symbiotic chloroplasts

Species	Plastid Source	References
Aceteonia senestra Quatrefages	*Codium fragile*	Taylor, 1967
Elysia atroviridis Baba	*Bryopsis corticulans*	Kawaguti and Yamasu, 1965
E. hedgpethi Marcus	*Codium fragile*	Greene, 1970 a, b
E. viridis Montagu	*Codium tomentosum*	Taylor, 1967, 1968
Hermaea bifida Montagu	*Griffithsia* sp.	
	Delesseria sp.	
Limapontia capitata Müeller		Taylor, 1967
Oxynoe panamensis Pilsbry & Olsson	*Caulerpa sertularioides*	
Placida dendritica Alder & Hancock	*Codium tomentosum*	Taylor, 1967
	Codium fragile	Greene, 1970a
	Bryopsis corticulans	
Placobranchus ianthobapsus Gould		Greene, 1970a
Tridachia crispata Mörch	*Caulerpa racemosa* (?)	Trench, et al., 1969
Tridachiella diomedea Bergh	*Caulerpa sertularioides* (?)	Trench, et al., 1969

*This observation is still under investigation.

slit, the animal sucks out the liquid cell contents, and it is with the cell sap that the animals acquire their chloroplasts. Presumably, mitochondria, nuclei, and other cell particulates are ingested as well, but their fate in unknown. Once ingested, the chloroplasts are moved by ciliary currents into the tubules of the digestive gland where, apparently by phagocytosis, they are taken into the digestive cells. At this point the story becomes

interesting, for in many species the chloroplasts do not seem to be
digested, but instead remain fully functional for extended periods
of time.

The chloroplasts must be obtained anew by each generation of
slugs since plastids are not transmitted in the eggs (Kawaguti and
Yamasu 1965; Greene 1968). Therefore, this is not an hereditary
symbiotic condition such as that which occurs in many invertebrate
groups which harbor algal symbionts.

LONGEVITY OF CHLOROPLAST SYMBIOSES

Algal chloroplasts which have gained entry into the cells of
the sacoglossan digestive gland remain for varying periods of time
depending on the species involved. Thus, plastids associated with
Placida dendritica lose their functional capacity within 24 hrs of
ingestion (Taylor 1968; Greene and Muscatine 1972), while those in
other species of slugs may survive considerably longer. It has been
demonstrated that chloroplast symbionts in *Tridachia crispata,
Tridachiella diomedea* (Trench, Greene, and Bystrom 1969), and *Placobranchus ianthobapsus* (Greene 1970b) may last for more than a month
in the animal's cells. The ultimate fate of the chloroplasts is
still open to discussion (see Kawaguti and Yamasu 1965; Taylor 1968;
and Greene 1970b).

PHOTOSYNTHESIS BY SYMBIOTIC CHLOROPLASTS

Photosynthetic function in chloroplasts symbiotic with sacoglossans has been demonstrated both by fixation of $^{14}CO_2$ in the
light and dark (Taylor 1968; Trench, Greene, and Bystrom 1969;
Greene and Muscatine 1972) and by O_2 production under light and dark
conditions (Taylor 1971; Testerman, unpublished report; Trench,
Greene, and Bystrom 1969; Kawaguti 1941). In most cases studied to
date, animals containing chloroplasts fix more ^{14}C in the light than
in the dark, and most associations produce oxygen in the light as
well. These two lines of evidence indicate that the plastids in the
animal cells are still quite functional.

Photosynthetic products of some animal-chloroplast associations
have been studied and compared to photosynthetic products of intact
host algae (Greene and Muscatine 1972; Trench, Trench, and Muscatine
1972). In both studies it was found that chloroplasts in animals
produced more intermediary metabolites than did the intact plants,
but the latter produced far more lipid material. Following incubation of animals with $H^{14}CO_3$, in other studies, the mucus produced by
the slugs was highly radioactive (Trench, Trench, and Muscatine 1970;
Trench, Trench, and Muscatine 1972). The radioactivity resulted
from high quantities of ^{14}C-galactose and ^{14}C-glucose in the mucus

fraction, indicating that photosynthesis by the symbiotic chloro-plasts was an important source of carbohydrate for mucus synthesis.

By incubating chloroplasts isolated from the alga, *Codium frag-ile*, with $^{14}CO_2$, Greene (1970c) showed that a single compound, ten-tatively identified as glycolic acid, was selectively leaked to the suspending medium by the plastids. It is possible that this same compound is leaked to the animal cells in the symbiotic association. Leakage of glycolic acid from the chloroplasts represented about 16% of the total carbon fixed. Trench, Trench, and Muscatine (1972) reported a leakage rate of about 50% of the total fixed carbon from the chloroplasts to the host animal in *Tridachia* and *Tridachiella*. Aside from the role played in mucus synthesis, reduced carbon from photosynthesis is also converted into alcohol-insoluble moieties in the animals. An estimated 7 to 10% of the total carbon fixed ends up as protein in the animal (Trench, Trench, and Muscatine 1972). The metabolic requirements of sacoglossan slugs are still unknown, but it is intriguing to consider the role of the chloroplast sym-biont in their total nutrition.

SOME EXCEPTIONS TO THE RULE

In the course of studies on chloroplast-sacoglossan symbioses, a few notable exceptions have been encountered. There are a few sacoglossans that are apparently unable to enter into a symbiotic association with plastids from their algal food. One such organism is the bivalved sacoglossan, *Berthelinia chloris*, from the Gulf of California. Haxo (personal communication) reports that *Berthelinia* does not produce oxygen in the light, suggesting that photosynthesis does not occur. This fact is interesting since the animal feeds on *Caulerpa sertularioides* which supplies chloroplasts to another saco-glossan, and *Berthelinia* is grass green in color although much of the pigment is apparently of animal origin.

Another algae-feeding sacoglossan, *Hermaeina smithi*, also lacks symbiotic chloroplasts. The capacity of this species to fix $^{14}CO_2$ in the light was no greater than it was in the dark following 24 hrs starvation (Greene 1970a). Thus, the chloroplasts are probably rapidly destroyed by the animal. *Olea hansineensis* is another saco-glossan without chloroplasts in its digestive gland cells. In this case, however, it is not so surprising since *Olea* does not feed on algae at all, but rather on eggs of other opisthobranchs (Hurst 1967). *Olea hansineensis* may well be the only sacoglossan which does not feed on algae.

SOME PERSPECTIVES IN CHLOROPLAST SYMBIOSES

It has become evident that many, if not most, sacoglossan opisthobranchs are capable of entering into a symbiotic association with algal chloroplasts. Perhaps chloroplast symbiosis is a general phenomenon among the Sacoglossa (see Greene 1970a), but there are several questions of great interest which have yet to be answered in this regard.

Placobranchus ianthobapsus from Hawaii has never been observed feeding on an alga in the field (Greene 1970a). This might mean that once a *Placobranchus* veliger settles and metamorphoses, it feeds on the appropriate siphonaceaous alga and then moves up to the reef-flats. If this is true, it would indicate that all chloroplasts in the cells of *Placobranchus* were derived from those obtained in the original feeding, or their progeny. So far, it has not been established whether chloroplasts in the cells of any animal are capable of replication.

Another aspect of the symbiosis which should be examined in *Placobranchus* is whether the chloroplasts in the animal are capable of synthesizing their own chlorophyll pigments. Greene (1970b) showed that *Placobranchus* did not lose any chlorophyll pigment (on a per animal basis) after being starved for twenty-seven days. Starvation in this case simply controls against the acquisition of new chloroplasts. Therefore, since all the chlorophyll still remains, it would indicate that everything necessary for chlorophyll-turnover is present in the animal-chloroplast association. This is especially interesting in that in higher plants, two nuclear genes are required in the synthesis of chlorophyll (Kirk and Tilney-Bassett 1967). Trench and Smith (1970) concluded that plastids in *Elysia viridis* were incapable of synthesizing chlorophyll. In the latter case, perhaps, it is not so surprising since the plastids in *Elysia* are not as permanent in their association.

Finally, even though we know to a great extent what is happening to the reduced carbon leaked to the animals by their symbiotic chloroplasts, we still do not know what material (materials?) is translocated by the circulatory system. This problem is not easily solved due to the problems inherent in getting blood samples from small organisms which produce copious amounts of mucus when disturbed.

There is one more aspect of chloroplast symbionts which should be mentioned. The discovery of chloroplasts symbiotic in animal cells has revived the early theories (Schimper 1885; Mereschkowsky 1905) that chloroplasts of higher plants really were symbionts that at one time in geologic history were free-living organisms. Much information is now available on this topic (see reviews by Taylor 1970; Schnepf and Brown 1971) and more is being added daily. One could conclude that if a chloroplast can be a symbiont in an animal cell, it should be able to do the same in a plant cell.

LITERATURE CITED

De Negri, A., and De Negri, G. 1876. Farbstoff aus *Elysia viridis*. Ber. Deut. chem. Gesellsch. 9:84.
Fretter, V. 1941. On the structure of the gut of the sacoglossan nudibranchs. Proc. Zool. Soc. Lond. Ser. b. 110:185-98.

Greene, R. W. 1968. The egg masses and veligers of southern
California sacoglossan opisthobranchs. Veliger 11:100-104.
_____. 1970a. Symbiosis in sacoglossan opisthobranchs: symbiosis
with algal chloroplasts. Malacologia 10:357-68.
_____. 1970b. Symbiosis in sacoglossan opisthobranchs: functional
capacity of symbiotic chloroplasts. Mar. Biol. 7:138-42.
_____. 1970c. Symbiosis in sacoglossan opisthobranchs: transloca-
tion of photosynthetic products from chloroplast to host tissue. Malacologia 10:369-80.
_____, and Muscatine, L. 1972. Symbiosis in sacoglossan opistho-
branchs: photosynthetic products of animal-chloroplast associations. Mar. Biol. 14:253-59.
Hurst, A. 1967. The egg masses and veligers of thirty Northeast
Pacific opisthobranchs. Veliger 9:255-88.
Kawaguti, S. 1941. Study on the invertebrates associating unicellular algae. I. *Placobranchus ocellatus* van Hasselt, a nudibranch. Palao Trop. Biol. Sta. Studies 2:307-308.
_____, Yamamoto, M., and Kamishima, Y. 1965. Electron microscopy
on the symbiosis between blue-green algae and an opisthobranch
Placobranchus. Proc. Japan Acad. 41:614-17.
_____; and Yamasu, T. 1965. Electron microscopy on the symbiosis
between an elysioid gastropod and chloroplasts of a green alga.
Biol. J. Okayama Univ. 11:57-65.
Kay, E. A. 1968. A review of the bivalved gastropods and a dis-
cussion of evolution within the Sacoglossa. Symp. Zool. Soc.
Lond. 22:109-34.
Kirk, J. T. O., and Tilney-Bassett, R. A. E. 1967. The plastids-
their chemistry, structure, growth and inheritance. San
Francisco: W. H. Freeman and Co.
Mereschkowsky, C. 1905. Über Natur und Ursprung der Chromatophoren
im Pflanzenreiche. Biol. Zbl. 25:593-604.
Ostergaard, J. M. 1955. Some opisthobranchiate Mollusca from
Hawaii. Pac. Sci. 9:110-36.
Schimper, A. T. W. 1885. Untersuchungen über die Chlorophyllkörner
und die ihnen homologen Gebilde. Jb. wiss. Botan. 16:1-247.
Schnepf, E., and Brown, R. M., Jr. 1971. On relationships between
endosymbiosis and the origin of plastids and mitochondria. In
ed. J. Reinert and H. Ursprung, Origin and continuity of cell
organelles. Cell Differentiation 2:299-322. New York:
Springer-Verlag.
Taylor, D. L. 1967. The occurrence and significance of endosymbiotic chloroplasts in the digestive glands of herbivorous opisthobranchs. J. Phycology 3:234-35.
_____. 1968. Chloroplasts as symbiotic organelles in the digestive
gland of *Elysia viridis* (Gastropoda: Opisthobranchia). J. mar.
biol. Ass. U. K. 48:1-15.
_____. 1970. Chloroplasts as symbiotic organelles. Int. Rev.
Cytol. 27:29-64.
_____. 1971. Photosynthesis of symbiotic chloroplasts in *Tridachia
crispata* (Bergh). Comp. Biochem. Physiol. 38A:233-36.

Trench, M. E., Trench, R. K., and Muscatine, L. 1970. Utilization of photosynthetic products of symbiotic chloroplasts in mucus synthesis by *Placobranchus ianthobapsus* (Gould), Opisthobranchia, Sacoglossa. Comp. Biochem. Physiol. 37:113-17.

Trench, R. K. 1969. Chloroplasts as functional endosymbionts in the mollusc *Tridachia crispata* (Bergh), (Opisthobranchia, Sacoglossa). Nature 222:1071-72.

_____, Greene, R. W., and Bystron, B. G. 1969. Chloroplasts as functional organelles in animal tissues. J. Cell Biol. 42:404-17.

_____, and Smith, D. C. 1970. Synthesis of pigment in symbiotic chloroplasts. Nature 227:196-97.

_____, Trench, M. E., and Muscatine, L. 1972. Symbiotic chloroplasts; their photosynthetic products and contribution to mucus synthesis in two marine slugs. Biol. Bull. 142:335-49.

Yonge, C. M., and Nicholas, H. M. 1940. Structure and function of the gut and symbiosis with zooxanthellaw in *Tridachia crispata* (Oerst.). Bgh. Pap. Tortugas Lab. 32:287-301.

Intraspecific aggression and the distribution of a symbiotic polychaete on its hosts

Ronald V. Dimock, Jr.
Department of Biology
Wake Forest University
Winston-Salem, North Carolina

One of the important characteristics of any population of living organisms is the dispersion, or spatial pattern of the individuals. The individuals of a population may exhibit a clumped or aggregated configuration, a uniform distribution, or a random spatial pattern. The results of numerous analyses of dispersion, principally in plants, suggest that few populations exhibit a completely random distribution (Greig-Smith 1964). A nonrandom distribution indicates that some kind of constraint is operating on that population to effect the observed distribution (Connell 1963). Since experimental evidence is usually lacking, the nature of the constraint generally must be inferred from the pattern itself. Clumping, for example, may be the result of gregarious behavior, reproductive habits, or heterogeneity of the environment with individuals being clumped in the favorable parts of the habitat. A uniform dispersion suggests that some negative interaction occurs between the individuals (Connell 1963).

Relatively few marine symbiotic associations have been systematically examined with regard to the nature of the spatial arrangement of a symbiont upon its hosts. All too frequently observations have been based on collections of inadequate sample size, have not considered possible seasonal variations, and have relied on collection techniques such as dredging, which may have resulted in the loss of some symbionts. In addition, usually the data have not been analyzed statistically to determine the actual spatial arrangement expressed. However, it seems clear that several species of symbiotic organisms are aggregated on their hosts. This seems to be true for various montacutid bivalves which occur in associations with a variety of hosts and may attain densities of up to 30 calms per host (Gage 1966). Similarly, up to 23 *Arthritica bifurca*, another small bivalve, have been observed associated with tubes of the polychaete *Pectinaria australis* (Wear 1966).

Aggregations of this sort are not limited to molluscs. As many as 370 of the copepod *Myocheres major* have been recorded from a single bivalve host (Humes and Cressey 1960). Similarly, 648 *Histriobdella homari* (Polychaeta) have been recovered from a single American lobster (Simon 1967). Many examples of symbionts occurring in aggregations could be listed and it seems pointless to document the plethora of such observations. Various groups of symbionts exhibit spatial patterns of their distributions which indicate clumping and thus resemble the pattern which is taken by several authors to be characteristic of many natural populations (Allee et al.1949; Greig-Smith 1964; Odum 1971).

In contrast to the occurrence of aggregated organisms, uniform distributions seem to occur far less frequently in nature (Odum 1971). Not surprisingly, various species of symbionts, primarily polychaetes and pinnotherid crabs, are numbered among those organisms which exhibit uniform distributions. These patterns frequently emerge only among adult symbionts, with juveniles randomly distributed upon their hosts (Stauber 1945; Christensen and McDermott 1958; Pearce 1966; Palmer 1968). The development of regularity in spatial distribution from a once random pattern implies that some mechanism regulates the distribution of these large or mature symbionts upon their hosts. Several mechanisms have been advanced to explain such distributions, e.g., inhibition of the growth of young by adults (Pearce 1964).

Aggressive behavior characteristic of territoriality has previously been reported for several annelids. The polychaete *Nereis pelagica* exhibits a fighting reaction which may be linked to tube defense (Clark 1959). Sex-specific fighting (i.e., male versus male, female versus female) has been reported for the polychaete *Neanthes caudata* (Reish 1957). Similarly, the polynoid *Hesperonoe* drives all intruders from the burrow it shares with *Urechis* (MacGinitie and MacGinitie 1968). Whether intraspecific aggression may influence the distribution of a symbiont upon its hosts has previously been the subject of speculation (Palmer 1968). Recent evidence (Goerke 1971) suggests that, indeed, such aggressive interactions may play a vital role in determining the distribution of the symbiotic polychaete *Nereis fucata*. In this paper field data are presented which clearly indicate that adults of the polychaete *Arctonoe pulchra* occur in a very regular pattern of distribution upon two of the species which serve as hosts for this symbiont. In addition, experimental evidence is presented which suggests that this pattern is at least in part influenced by the intraspecific aggression exhibited by these worms.

MATERIALS AND METHODS

Field Studies

Populations of the symbiotic polychaete A. *pulchra* were extensively sampled from two host species, the sea cucumber *Stichopus
parvimensis* and the limpet *Megathura crenulata*. These host organisms were collected from Naples Reef, a subtidal reef approximately
20 km west of Santa Barbara, California. For the censuses of worm
distribution, the hosts were gathered by hand by diving and placed
in individual plastic bags immediately upon removal from the substrate to prevent the loss of any symbionts. The individually bag-
ged hosts were brought to the laboratory where they were held overnight (usually 20-24 hrs). This treatment usually resulted in nar-
cotizing both the hosts and their symbionts, since the oxygen in
the bags reached low levels. The hosts and the contents of the bags
were subsequently examined for the presence of worms and care was
taken to insure as accurate a census as possible. Worms were measured to the nearest millimeter. No attempt was made to measure the
hosts. Samples of approximately thirty *Stichopus* were collected
monthly from December 1968 through March 1970 except during January,
February, and December 1969. Similar collections of *Megathura* were
made from September 1969 through March 1970 except during October
and December 1969.

Experimental Studies

A series of laboratory investigations was performed to determine whether *Arctonoe* could regulate its density on the sea cucumber
host, *S. parvimensis*. In these experiments all references to hosts
refer to this sea cucumber. All of the worms used in these studies
were large A. *pulchra*, i.e., worms greater than 20 mm in length.
The experimental designs will be described here. Specific variations will be indicated in the appropriate sections of the results.
Two kinds of experiments were performed in which the density of
the worms on their hosts was artificially increased, i.e., manipu-
lated by the investigator. In the first experiment the host sea
cucumbers were free to move about the experimental aquarium. In the
remaining studies the hosts were restrained by being placed under
inverted plastic vegetable collanders which were elevated on three
1/4-in wooden blocks and anchored by a brick. This arrangement permitted the manipulation of the hosts into whatever array was appropriate to the experiment in question and effectively prevented
physical contact between any two hosts. All experiments were conducted in an aquarium (1 x 1.6 x .18 m) which was supplied with
seawater inlets at two opposite corners and equipped with a drain
which was slightly off center. This indoor aquarium was subjected

to a slightly irregular light schedule since the room was occasionally used for short periods in the evenings; otherwise, diurnal illumination prevailed.

In these experiments the density of the worms on this host was varied up to a maximum of an initial density of five per host. The distribution of these worms was subsequently monitored at various time intervals. Care was taken not to handle the hosts excessively at each census. However, as will be noted below, the worms frequently are hidden from the view of an observer and some of the worms could not be accounted for during particular censuses in the course of certain experiments. All of these investigations were conducted at the Marine Laboratory of the University of California, Santa Barbara, California.

RESULTS

Distribution on Two Species of Hosts: Field Data

During the course of the field collections, 625 worms were recovered from 384 sea cucumbers and 72 worms from 148 limpets. The distribution of these worms on their respective hosts varied seasonally. A summary of the overall incidence of occurrence of *A. pulchra* on these two species of hosts is presented in Figure 1. It is apparent that the two species examined in this study do not serve equally as hosts for this symbiont. The incidence of occurrence of *A. pulchra* on *S. parvimensis* is consistently greater than that on the sympatric host *M. crenulata*. These data suggest that fewer host-symbiont associations occur between *Arctonoe* and *Megathura* than occur between *Arctonoe* and *Stichopus*. However, nothing can be said at present about differences that might exist in the absolute number of host-symbiont associations within these two host populations as no data are available concerning the total numbers of each species of host that occur in this collection area. The data from both host populations, however, clearly indicate a pronounced seasonal variation in the frequencies with which these two species of host support populations of this symbiont. The occurrence of *A. pulchra* varied from a high of 100% infestation of *Stichopus* during the summer months to a low of 60% in February and from a high of 80% to a low of 25% on *Megathura* during comparable periods.

The data concerning the incidence of occurrence of *Arctonoe* on these two hosts reveal something of the dynamics of the interactions of these specific host-symbiont associations. Of considerably greater interest is the pattern of the distribution of these worms upon their hosts once an association has been effected. An analysis of the distribution of *A. pulchra* on *S. parvimensis* will serve to illustrate this fact.

The data in Table 1 show that multiple colonizations of the sea cucumber host occur frequently, particularly during the summer

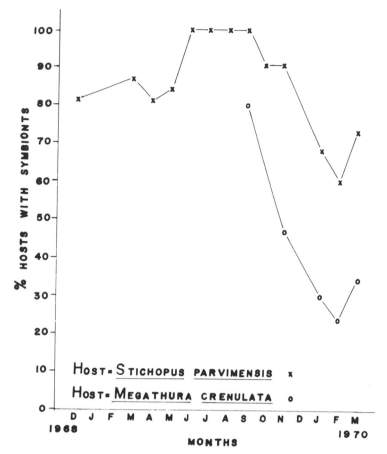

Figure 1. Incidence of occurrence of *Arctonoe pulchra* on its hosts.

months. This is evident from the fact that of the 329 sea cucumbers
having worms associated with them, 169 had two or more, at least
when all size classes were considered (Table 1). These data also
reflect a significant deviation from random (toward uniformity) in
the spatial distribution of this population of worms on this host
(analysis in Table 1).

The data summarized in Table 1 can be partitioned with respect
to two general size classes of worms recovered from this host. When
this is done, a very interesting pattern emerges (Tables 2 and 3).
It is immediately apparent that the distribution of the large worms
(>20 mm)(Table 3) is very different from that of the small worms
(Table 2). Whereas the distribution of the small worms (<20 mm)
is for the most part completely random on this host (Table 2), the
large worms are uniformly distributed and occur in a very regular
pattern of isolation of a single large worm per host (Table 3).

A similar trend, though admittedly not as obvious, also per-
tains in the distribution of *A. pulchra* on the limpet *M. crenulata*
(Table 4). Again, no cases were observed in which two or more large
worms occurred simultaneously on a given limpet. However, the uniformity of the dispersion is not as clear-cut as it is on the sea

TABLE 1
Observed densities of *Arctonoe pulchra* of all size classes on
Stichopus parvimensis

No./Host	\multicolumn{13}{c}{Number of hosts examined having the densities shown}
	Dec 68	Mar 69	Apr 69	May 69	Jun 69	Jul 69	Aug 69	Sep 69	Oct 69	Nov 69	Jan 70	Feb 70	Mar 70
0	6	4	6	5	0	0	0	0	2	2	10	12	8
1	20	19	21	17	3	11	3	5	11	15	20	15	11
2	5	5	4	8	5	10	2	10	9	10	1	3	7
3		3		1	7	5	8	10	7	3			4
4					3	3	5	3	1				
5						1	5	1					
6					2		3	1					
7							3						
N	31	31	31	31	20	30	29	30	30	30	31	30	30
$\bar{X}$	.97	1.23	.94	1.16	2.90	2.10	3.97	2.60	1.80	1.47	.71	.70	1.23
s^2	.37	.65	.33	.54	1.99	1.27	3.11	1.42	.99	.60	.28	.42	1.01
$\dfrac{s^2}{\bar{X}}$	.38[*]	.53[*]	.35[*]	.47[*]	.69	.60[*]	.78	.55[*]	.55[*]	.41[*]	.39[*]	.60[*]	.82

[*]Indicates a deviation from random as calculated from $X^2 = s^2(N-1) \div \bar{X}$, $P < 0.05$ (Southwood, 1966, p. 36).

cucumber host (Table 3). This is very likely a reflection of the sparseness of the population of *A. pulchra* that occurs on this host (Southwood 1966, p.33).

Before leaving the field data concerning the distribution of this symbiont, some peripheral observations should be presented. When one observes *S. parvimensis* in the field, one readily notes that large worms are rarely visible anywhere on the exterior of these sea cucumbers. Although these observations have not been quantified, it appears that fewer large worms may be observed on the surface of these hosts in the field than are actually present (as shown by the worm census data). Also, during some preliminary experiments involving an analysis of the effects of a light:dark cycle upon the behavior of these worms on this host, worms consistently were found during the light phase of the cycle out of sight either in the oral cavity or on the sole or underside of the sea cucumber, but during the dark phase of the cycle frequently they were in sight on the upper, exposed surface of the host. Furthermore, in the laboratory worms were frequently observed in the oral cavity of *Stichopus*, either among the oral tentacles or actually backed into the esophagus so that only the head of the worm was visible in the oral cavity. At no time were two large worms found simultaneously in this region. The possible significance of these observations will be considered below.

TABLE 2
Observed densities of *A. pulchra* $\bar{}$ < 20 mm on *S. parvimensis*

No./Host	Number of hosts examined having the densities shown												
	Dec 68	Mar 69	Apr 69	May 69	Jun 69	Jul 69	Aug 69	Sep 69	Oct 69	Nov 69	Jan 70	Feb 70	Mar 70
0	20	16	15	13	2	3	0	2	11	11	26	23	15
1	11	12	13	15	4	11	4	10	2	2	5	5	10
2		3	3	2	7	8	6	10	11	15		2	4
3				1	4	6	7	5	9	10			1
4					1	2	3	2	7	3			
5					1		6	1	1				
6					1		2						
7							1						
N	31	31	31	31	20	30	29	30	30	30	31	30	30
$\bar{X}$	.35	.58	.61	.70	2.25	1.76	3.37	1.93	2.80	2.38	.16	.30	.70
s^2	.23	.45	.44	.54	2.30	1.21	2.88	1.37	.99	.60	.14	.35	.70
$\dfrac{s^2}{\bar{X}}$	.65	.77	.72	.77	1.02	.68	.85	.70	.35*	.25*	.87	1.16	1.00

*Indicates a deviation from random as calculated from $\chi^2 = s^2(N-1) \div \bar{X}$, $P < 0.05$ (Southwood, 1966, p. 36).

TABLE 3
Observed densities of *A. pulchra* > 20 mm on *S. parvimensis*

No./Host	Number of hosts examined having the densities shown												
	Dec 68	Mar 69	Apr 69	May 69	Jun 69	Jul 69	Aug 69	Sep 69	Oct 69	Nov 69	Jan 70	Feb 70	Mar 70
0	12	12	21	17	8	20	13	12	7	10	14	18	14
1	19	18	10	14	11	10	15	16	22	20	17	12	16
2		1			1		1	2	1				
N	31	31	31	31	20	30	29	30	30	30	31	30	30
$\bar{X}$	.61	.64	.32	.45	.65	.33	.55	.67	.80	.67	.54	.40	.53
s^2	.25	.30	.22	.25	.34	.22	.32	.36	.23	.23	.26	.24	.25
$\dfrac{s^2}{\bar{X}}$	.41*	.46*	.68	.55*	.52*	.66	.58*	.53*	.28*	.34*	.48*	.60*	.47*

*Indicates a deviation from random as calculated from $\chi^2 = s^2(N-1) \div \bar{X}$, $P < 0.05$ (Southwood, 1966, p. 36).

TABLE 4
Observed densities of *A. pulchra* on *Megathura crenulata*

No./Host	Number of hosts examined having the densities shown									
	All size classes of worms					Worms > 20 mm				
	Sep 69	Nov 69	Jan 70	Feb 70	Mar 70	Sep 69	Nov 69	Jan 70	Feb 70	Mar 70
0	6	16	21	22	19	10	16	22	22	20
1	20	14	9	7	10	20	14	8	7	9
2	1									
3	2									
4	1									
N	30	30	30	29	29	30	30	30	29	29
$\bar{X}$	1.06	.47	.30	.23	.33	.67	.47	.27	.23	.31
s^2	1.18	.25	.22	.19	.24	.23	.25	.21	.19	.22
$\frac{s^2}{\bar{X}}$	1.11	.53*	.73	.82	.73	.34*	.53*	.78	.82	.71

*Indicates a deviation from random as calculated from $\chi^2 = s^2(N-1) \div \bar{X}$, $P < 0.05$ (Southwood, 1966, p. 36).

Intraspecific Aggression

Early in this study it became very apparent that *Arctonoe* is very aggressive toward conspecifics. Worms which were placed in a common container invariably bit one another with their armed proboscises. Frequently, this behavior produced open wounds and torn appendages among the worms in a container. Several encounters were witnessed in which complete segments were bitten from a worm, or small worms were ingested by larger worms.

This aggressive behavior is not limited to worms which have been removed from their hosts. Worms which were maintained on their hosts for long periods but in crowded conditions within an aquarium often exhibited wounds and missing appendages. Furthermore, when two worms approached each other on the surface of a host, they would frequently jerk away on contact. Occasionally such contact resulted in biting. Several interactions were observed in which one or both of the *worms actually left the host they were on at the time of the aggressive encounter and moved onto an adjacent host.*

Although no detailed records were kept, aggression occurred among most of the size classes of worms. Very small worms were never observed to bite the largest worms, but interactions among various size combinations took place. There were no noticeable differences in behavior which could be attributed to differences in the sex of the worms. It would seem probable that such aggressive behavior might in part influence the ultimate spatial distribution of *A. pulchra* on its hosts.

Relocation Experiments: *Arctonoe* on *Stichopus*

Aggressive interactions among these worms could affect the distribution of large *Arctonoe* if such interactions: (1) resulted in the deaths of some of the worms, or (2) stimulated one or more worms to leave a host, regardless of the subsequent fate of those worms. If such interactions result in a worm's leaving a particular host, increasing the density of large worms on a host (presumably increasing the opportunity for aggressive encounters) might stimulate relocation behavior among some of the symbionts. The following series of experiments were performed to test this hypothesis.

Experiment 1. Twenty-five large worms were distributed among twenty-five sea cucumbers (previously freed of all worms) such that five hosts had five worms each and twenty hosts had no worms. The hosts were free to move about the experimental aquarium and frequently had physical contact with one another. The distribution of the worms on these hosts was subsequently noted at irregular intervals. The results of this experiment are depicted in Figure 2.

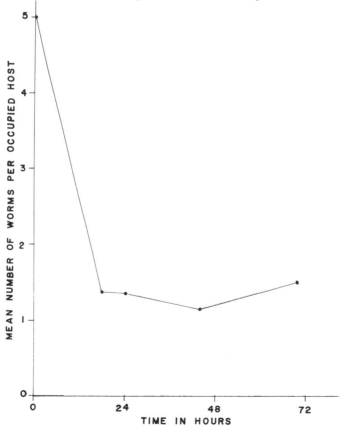

Figure 2. Density regulation by large *Arctonoe pulchra* on *Stichopus parvimensis:* hosts unrestrained. At the start, twenty hosts had no worms and five hosts had five worms each.

It is obvious from the data in Figure 2 that the worms did in fact rapidly redistribute themselves among the hosts. In less than 24 hrs the worms had occupied eighteen of the twenty-five hosts. Since three or four worms were unaccounted for at each census (though all were found at the termination of the experiment), additional hosts scored as having no worm might have been occupied, thus indicating even greater relocation activity of these symbionts. The relocation behavior of the worms lowered their density upon these hosts to a level very close to that found in nature, namely, one large worm per host. In a control experiment in which the initial density of twenty-five worms was one per host, no instances were observed during a similar time period of more than two worms occupying a single host at a given time, and these double occupancies were rare occurrences.

Experiment 2. Twenty-five large worms were again distributed among twenty-five sea cucumbers such that five hosts had five worms each and twenty hosts had no worms. Unlike the situation in Experiment 1, in this experiment the hosts were restrained (see Materials and Methods) and no physical contact could occur between hosts. Thus, for any change in density to occur, a worm or worms would have to leave a host, and to effect contact with another host would require that the worm traverse several inches of the bottom of the tank.

In view of the conditions imposed on these worms, the resulting changes in density which did occur are remarkable (Fig. 3). Substantial relocation occurred. Eight previously unoccupied hosts had at least one worm on them within 24 hours following the beginning of this experiment. All twenty-five worms were accounted for at the first census (24 hrs), but only twenty-two and twenty-one were observed at the second and third censuses, respectively. Thus, even greater relocation may have occurred than can be interpreted from these data.

Circumstantial evidence indicated that much aggression had occurred during these experiments. Many of the worms were badly injured. The unusually high density of worms, particularly in Experiment 2 in which a worm had no choice but to leave a host to find additional "room," perhaps resulted in the worms' sustaining numerous severe injuries before leaving a host. These injuries, perhaps more numerous and more severe than they would have been at a lower density of worms, may have prevented more worms from seeking refuge by leaving a host. In addition, the occasional occurrence of double occupancy by large worms observed in the field census may indicate that these worms would be more tolerant of one additional worm than they are of more than one additional conspecific on a given host. Therefore, one additional experiment utilizing a lower initial worm density was performed.

Experiment 3. Seven hosts, each with one large worm which had been on it at the time of collection, were arranged along with eight hosts, previously freed of worms, in an array of three rows in the experimental tank. None of the worms which were present moved from its host during the next eight days (Table 5), despite the fact that

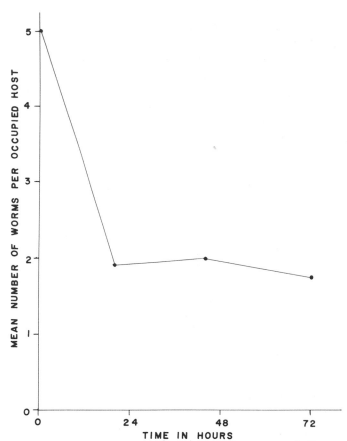

Figure 3. Density regulation by large *Arctonoe pulchra* on *Stichopus parvimensis*: host restrained. At the start twenty hosts had no worms and five hosts had five worms each.

TABLE 5
Regulation of density by large *Arctonoe pulchra*

Time in Days	Individual Hosts														
	A	B	C	D	E	F	G	H	I	J	K	L	M	N	O
0	0	1	0	1	0	1	0	1	0	1	0	1	0	1	0
5 worms rotated	1	0	1	0	1	0	1	0	1	0	1	0	1	0	0
8 5 worms added	1	0	2	0	2	0	2	0	2	0	2	0	1	0	0
9	1	1	2	1	1	1	1	0	1	0	1	0	2	0	0
12	1	1	1	1	1	1	1	1	1	0	1	0	1	1	0

*Entries are the number of worms on individual *Stichopus parvimensis*.

each of the seven worms was moved to a different sea cucumber on the fifth day (as a control for any effect of manipulation by the experimenter on the worm's behavior). On the eighth day of this experiment five additional worms were introduced onto the array of hosts so that five hosts had two worms each, two hosts had one worm each, and eight hosts remained free of any worms. The distribution of the twelve worms among the fifteen hosts was then monitored over four days.

The remarkable resulting distributions (Table 5) were as follows. Within 24 hours four of the ten worms which had been present as pairs on their hosts had moved to another host. Of these four, all but one moved to an unoccupied host. The fourth moved onto an already occupied host. Thus, within 24 hrs the worms had redistributed themselves so that eight hosts had one worm each and two hosts each had two worms on them. The distribution was not checked for three days, but at that time all twelve worms were distributed singly on twelve of the fifteen hosts.

DISCUSSION

Associations between the symbiotic polychaete *A. pulchra* and two of its hosts in Southern California, *S. parvimensis* and *M. crenulata*, occur frequently, particularly during the summer months. Multiple colonization of both hosts occurs. However, the pattern of distribution of these worms upon their hosts varies as the worms become larger. Small, presumably young, worms are essentially randomly dispersed on their hosts. In contrast to this pattern the large worms are uniformly distributed on their hosts and occur as isolated single individuals. Such isolation of the large individuals may be rather typical of the distribution of several symbiotic polynoids, since such a pattern occurs in the association of *Lepidonotus sublevis* and its hermit crab host *Pagurus pollicaris* at Woods Hole, Massachusetts (unpublished observations) and with *Arctonoe vittata* and the limpet *Diodora aspera* in Oregon (Palmer 1968). Numerous casual observations in the literature (e.g., Davenport 1953) also suggest that the occurrence of a single scale worm upon its host is a very common phenomenon.

What could be the significance of this uniform isolation of the large worms? Intuitively, one would think that some other configuration such as an isolation of bisexual pairs of *A. pulchra* on their hosts might be biologically more sound than the observed distribution of one worm per host. Such a pattern has been observed for several other symbionts (Gray 1961; Pearce 1966) and would seem to be reproductively advantageous as males and females then would be in closer proximity to one another. However, subjective evaluation of the distributions of the hosts for this symbiont suggests that the host populations are themselves clumped rather than random or uniform. Such a spatial arrangement of the hosts in any local habitat would insure that the symbionts would be close enough to one another so that with some mechanism to synchronize gamete release, reproductive success would be more likely to occur than one might infer from the observed isolation of the worms upon their hosts.

Perhaps this regular isolation may confer some reproductive advantage to the symbionts, as has been postulated for some bird territories (Hinde 1956; Armstrong 1965). Density regulation may exclude some members from the breeding population and this may be of selective advantage. The fact that most of these worms are sexually mature at a length of 20 mm or so, the size when the pattern of regular isolation appears in their distribution, might indicate an association between maturity and this distribution.

The selective advantages enjoyed by a species living in a symbiotic association, though at present subject to theoretical considerations only, might be intimately associated with determining the optimum density of that symbiont upon its hosts. If *Arctonoe* were to profit from its associations with its hosts only by "having a place to live," one might readily expect that more than one large worm could live with each host simply by virtue of the relatively large surface area that the host represents. However, if the worms are dependent upon the host for more than merely a substrate on which to live, then a more restricted portion of the "host space" might be necessary for the worm's survival. The observations regarding the frequent occupation of the oral cavity of *Stichopus* by *A. pulchra* suggest that this worm might indeed spend much of its time on a rather restricted portion of the host's surface, namely, the oral region. Such behavior might be of adaptive significance to this worm since this is an ideal location for a symbiont to enjoy protection and at the same time be provided with a rich food supply. It could be hypothesized that these worms "defend" this oral cavity or keep their density at one large worm per host to insure that this region be available for occupancy at all times.

It should be possible to determine the amount of food available to a symbiont living on *Stichopus* and the amount of food required by *Arctonoe*. Such considerations have recently been attempted for another symbiotic association (Castro 1971). From such information it might be possible to determine whether a successful symbiont must spend much of its time in this oral cavity to get an adequate food supply. If this is indeed the case, then one selective advantage to the observed distribution would be that it reduces competition for a scarce resource.

Regardless of the significance of the distribution of large *A. pulchra*, the development of a uniform distribution from a random configuration suggests that some constraint operates upon the populations to effect this regular isolation of large worms. Evidence from this study shows that intraspecific aggression may serve as such a constraint. Clearly, the presence of one or more large *A. pulchra* on a host sea cucumber affected the behavior of additional large worms on that host. Circumstantial evidence indicated that this effect was mediated by aggressive interactions, interactions which result in the relocation of some of the worms.

Laboratory experiments demonstrated that worms could in some manner be stimulated to leave one host and move to another. Field data indicating movement among hosts by adult symbionts are few. However, Palmer (1968) has provided indirect evidence that such movements may in fact occur in a population of the congeneric polynoid *A. vittata*. In that study, host gastropods which had previously been freed of all worms were placed in the field for varying lengths of time. When these marked hosts were later recovered, several of them were found to have large worms on them, worms which could not have grown to the observed size on these hosts during the experiment and, therefore, must have come from other hosts or a free-living population. Since there is little evidence that free-living populations of that polychaete exist, it is assumed that these newly recruited worms came from other hosts. It is probable that similar experiments with hosts of *A. pulchra* would yield similar evidence of recruitment of large worms. The relatively rare occurrence of two large worms on a *Stichopus* might be an instance in which a large worm had recently arrived on a host already occupied by one large worm.

The laboratory demonstration of mobility within a population of *A. pulchra* and the field evidence which suggests that this mobility may in fact occur in nature indicate a possible role for the chemically mediated host recognition behavior exhibited by this worm (Dimock and Davenport 1971). Obligatory symbionts which for any reason become separated from their hosts must find another host or die. *A. pulchra* clearly is capable of recognizing some of its hosts from among an array of organisms (Dimock and Davenport 1971). This recognition behavior is quite specific for a particular species of host and the response continues for at least one month following separation of the worm and its host. Thus, a worm which has been stimulated to leave a host, perhaps as the result of an aggressive encounter with a conspecific, can utilize this specific recognition behavior to effect an association with another appropriate host. Specificity in this recognition behavior assures that the worm effects an association with the same species of host, a species to which it presumably is adapted, perhaps as the result of cryptic coloration (Dimock 1970). Therefore, this behavioral capability of *Arctonoe* is of fundamental ecological and evolutionary significance to this polychaete.

SUMMARY

The distribution of the symbiotic polychaete *A. pulchra* on two of its hosts in southern California, the sea cucumber *S. parvimensis* and the limpet *M. crenulata*, has been extensively examined. Multiple colonizations of these hosts by this worm occur frequently; however, the distribution of the large worms (>20 mm) is very regular at one worm per host. Behavioral studies in the laboratory indicate that intraspecific aggression within populations of this worm results in these worms distributing themselves in a manner which regulates their density toward that which occurs in the field, namely, one large worm per host. The significance of this behavior and of the pattern of distribution exhibited by these worms is discussed.

ACKNOWLEDGMENTS

This research was supported under ONR Contract No. 4-222(03), with Dr. Demorest Davenport as Principal Investigator.

The encouragement and support offered by Dr. Davenport through-out this study are gratefully acknowledged. Dr. Marian Pettibone of the U. S. National Museum verified the identifications of the polychaetes employed in this study. The stimulating discussions and assistance in many aspects of this project contributed by Drs. Barry Ache, Eldon Ball, and Bill Stewart proved to be invaluable. Additional thanks go to Dr. Robert Sullivan for a critical review of this manuscript.

LITERATURE CITED

Allee, W. C., Emerson, A. E., Park, O., Park, T., and Schmidt, K. P. 1949. Principles of animal ecology. Philadelphia: W. B. Saunders.

Armstrong, J. T. 1965. Breeding home range in the nighthawk and other birds; its evolutionary and ecological significance. Ecology 46:619-29.

Castro, P. 1971. Nutritional aspects of the symbiosis between Echinoecus pentagonus and its host in Hawaii, Echinothrix calamaris. In Aspects of the biology of symbiosis, ed. T. C. Cheng, pp. 229-47. Baltimore: University Park Press.

Christensen, A. M., and McDermott, J. J. 1958. Life-history and biology of the oyster crab, Pinnotheres ostreum Say. Biol.Bull. 114:146-79.

Clark, R. B. 1959. The tubicolous habit and the fighting reactions of the polychaete Nereis pelagica. Anim. Behav. 7:85-90.

Connell, J. H. 1963. Territorial behavior and dispersion in some marine invertebrates. Res. Popul. Ecol. 5:87-101.

Davenport, D. 1953. Studies in the physiology of commensalism. III. The polynoid genera Acholoe, Gattyana and Lepidasthenia. J. mar. biol. Ass. U. K. 32:161-73.

Dimock, R. V., Jr. 1970. Ecological and physiological aspects of host recognition by a symbiotic polychaete. Ph.D. thesis, University of California, Santa Barbara.

_____; and Davenport, D. 1971. Behavioral specificity and the induction of host recognition in a symbiotic polychaete. Biol. Bull. 141:472-84.

Gage, J. 1966. Observations on the bivalves *Montacuta substriata* and *M. ferruginosa*, 'commensals' with spatangoids. J. mar. biol. Ass. U. K. 46:49-70.

Goerke, H. 1971. *Nereis fucata* (Polychaeta, Nereidae) als Kommensale von *Eupagurus bernhardus* (Crustacea, Paguridae). Entwicklung einer Population und Verhalten der Art. Veroff. Inst. Meeresforsch. Bremerh. 13:79-118.

Gray, I. E. 1961. Changes in abundance of the commensal crabs of *Chaetopterus*. Biol. Bull. 120:353-59.

Greig-Smith, P. 1964. Quantitative plant ecology. 2nd ed. Washington: Butterworth.

Hinde, R. A. 1956. The biological significance of the territories of birds. Ibis 98:340-69.

Humes, A. G., and Cressey, R. F. 1960. Seasonal changes and host relationships of *Myocheres major* (Williams), a cyclopoid copepod from pelecypods. Crustaceana 1:307-25.

MacGinitie, G. E., and MacGinitie, N. 1968. Natural history of marine animals. 2nd ed. New York: McGraw-Hill.

Odum, E. P. 1971. Fundamentals of ecology. 3rd ed. Philadelphia: Saunders.

Palmer, J. B. 1968. An analysis of the distribution of a commensal polynoid on its hosts. Ph.D. thesis, University of Oregon.

Pearce, J. B. 1964. On reproduction in *Pinnotheres maculatus* (Decapoda: Pinnotheridae). Biol. Bull. 127:384.

_____. 1966. On *Pinnixa faba* and *Pinnixa littoralis* (Decapoda: Pinnotheridae) symbiotic with the clam, *Tresus capax* (Pelecypoda: Mactridae). In Some contemporary studies in marine science, ed. H. Barnes, pp. 565-89. London: George Allen and Unwin.

Reish, D. J. 1957. The life history of the polychaetous annelid *Neanthes caudata* (Delle Chiaje), including a summary of development in the family Nereidae. Pacific Sci. 11:216-88.

Simon, J. L. 1967. Behavioral aspects of *Histriobdella homari*, an annelid commensal of the American lobster. Biol. Bull. 133:450.

Southwood, T. R. E. 1966. Ecological methods. London: Methuen & Co. Ltd.

Stauber, L. A. 1945. *Pinnotheres ostreum*, parasitic on the American oyster, *Ostrea (Gryphaea) virginica*. Biol. Bull. 88:269-91.

Wear, R. G. 1966. Physiological and ecological studies on the bivalve mollusk *Arthritica bifurca* (Webster, 1908) living commensally with the tubicolous polychaete *Pectinaria australis* Ehlers, 1905. Biol. Bull. 130:141-49.

The experimental analysis of host location in symbiotic marine invertebrates

Barry W. Ache
Department of Biological Sciences
Florida Atlantic University
Boca Raton, Florida

Basic to understanding the biology of symbiosis is the question: How are the partners brought together to initiate or to maintain these associations? As was noted by Davenport in two earlier considerations of this problem (1955, 1966), the host represents a source of primary stimuli controlling the host-oriented behavior of a more mobile partner. The experimenter can readily manipulate this "package" of stimuli in both time and space to rigorously quantify the behavioral and physiological mechanisms serving to bring the partners together.

This paper summarizes experimental studies of host location in invertebrate symbioses that have appeared since the reviews of Davenport (1966) and Cheng (1967) and presents some of the author's own data on the sensory basis of host location in symbiotic caridean shrimp. In keeping with the nature of the symposium, the review and discussion focus on the symbiotic associations of marine organisms. This is not to discredit the equally interesting advances that have been made toward understanding host location in nonmarine symbioses, particularly the freshwater snail-trematode miracidia relationships (see reviews of Cheng 1967; Ulmer 1970). Also relevant to a generalized discussion of host location, but not considered in the present paper are the terrestrial pollination associations (e.g., van der Pijl and Dodson 1966).

The process of host finding or host location could simply be a matter of random movement by a symbiont until its termination by a chance encounter with an appropriate host. The primary contribution of the earlier studies of host-symbiont interactions, however, was that many symbiotic organisms are behaviorally competent to detect and orient toward sources of stimuli at least partially of host origin, generating the idea that host location involves an active, overt response of the symbiont toward the host. Experimentally, one can consider an initial period of "distant" host recognition with the symbiont at some distance from the host followed by or concurrent with a period of oriented locomotion in which the symbiont moves toward the host. Once host contact is established, there may also be a period of "contact" host recognition that serves to complete or reinforce "distant" host recognition. Each of these events may be controlled by a unique set of stimuli or, as has been suggested for nudibranch-coelenterate feeding relationships (Harris 1971), by varied parameters of a single stimulus modality. Most experimental analyses, however, use oriented locomotion toward the host as a behavioral assay system, making it difficult, without further experimentation, to resolve the stimuli mediating host recognition from those mediating oriented locomotion. In the present discussion, the term host location is used collectively to include both discrimination of and orientation toward the host. Host location in natural situations may encompass an even broader sequence of events in which the symbiont initially responds to a stimulus or series of stimuli that characterize the general habitat of the host prior to responding to stimuli of host origin. Such a mechanism has been suggested for host location by trematode miracidia (Wright 1959) and hymenopteran parasites (Laing 1937) as well as for marine symbionts (Gage 1968).

HOST LOCATION VERSUS LIFE CYCLE

One can assume symbionts with planktonic larval stages must reassociate with an appropriate host at least once in each generation. Investigation with larval or recently settled symbionts, however, has been limited severely by the difficulty in rearing the numbers of healthy larvae required for behavioral studies. Consequently, most studies focus on the behavior of adult symbionts serving to maintain established relationships. Castro (1969) was successful in rearing a small number of the parthenopid crab, *Echinoecus pentagonus*, associated with the Hawaiian sea urchin, *Echinothrix calamaris*. Megalopae spend significantly more time in contact with freshly removed spines of the host than with similar but washed spines; this suggests chemical orientation might be an active component of the initial host-finding behavior. Gage (1966b) has pursued more detailed life history studies of the erycinacean bivalves *Montacuta substriata* and *M. ferruginosa* symbiotic with echinoids. Both laboratory-reared and field-collected larvae, being more active than adult bivalves, exhibit walking movements in response to the same stimulus parameters as do the adult symbionts. Gage suggests that the adult competence for host location represents a retention of larval behavior. In support of this idea is Gilpin-Brown's (1969) observation that both recently settled and adult *Nereis fucata* respond with similar searching movements to stimulation with low frequency substrate vibrations produced by movements of their host hermit crabs across the substrate, although the response of adults is less rigorous. Gage's hypothesis, if applicable to other partnerships, provides a way to interpret experimental results obtained with adult symbionts relative to the initiation of symbiotic relationships (but see the following discussion of conditioning).

Some recent evidence suggests, however, that <u>adult</u> symbionts actively relocate their hosts in order to maintain their partnerships. Dimock (see page 27) demonstrated that in the laboratory the polynoid worm, *Arctonoe pulchra*, redistributes itself among specimens of the host, the holothuroid *Stichopus parvimensis*, confined in cages approximately 10-15 cm apart within a period of 24 hrs. Laboratory observations by the author suggest that two species of shrimp, *Betaeus harfordi* and *B. macginitieae*, symbiotic respectively with the abalone, *Haliotis rufescens*, and the sea urchin, *Strongylocentrotus franciscanus*, regularly leave the protection of their hosts during periods of laboratory-imposed darkness, returning within minutes of the onset of light.* To quantify this behavior, specimens of *B. macginitieae* or *B. harfordi* were observed with their respective hosts in a large sea-table maintained under a 12 hr light-12 hr dark regime. Table 1 summarizes the results obtained by observing the location of the shrimps relative to their hosts once during each

TABLE 1

Light/dark activity experiments: Betaeid shrimps associated with sea urchins and abalone

No. Hosts	No. Shrimp	Total No. shrimp exposed		No. L:D cycles
		Light	Dark	
B. macginitieae				
5[a]	2	0	5	5
5	5	0	14	5
5	10	0	23	4
B. harfordi				
4[b]	2	0	3	4
4	4	0	10	4
4	8	0	31	5

[a] 5 *Strongylocentrotus franciscanus*.
[b] 4 *Haliotis rufescens*.

*This rapid host location response provided an excellent opportunity to quantify the stimuli mediating host location and, as discussed below, was developed as an assay for more detailed investigation of the *Betaeus-Haliotis* and the *Betaeus-Strongylocentrotus* associations.

light and dark period at three different densities of shrimps:hosts.
During light periods, shrimps of both species remained under or
immediately adjacent to their respective hosts. In the dark, shrimps
were observed up to 2.0 m away from the nearest host, the maximum
distance allowable in the holding situation. The data further suggest that not all shrimps were exposed throughout the dark periods.
In fact, hourly observations of *B. macginitieae* (five hosts:five
shrimps) during one complete 12 hr dark period indicated that the
number of individuals exposed varied from one to five. Gray,
McCloskey, and Weihe (1968) also report nocturnal wanderings in
laboratory-hold pinnotherid crabs associated with echinoids.

CHEMICALLY MEDIATED HOST LOCATION

As suggested by earlier studies (see reviews of Davenport 1966;
Cheng 1967; McCauley 1969), chemically mediated information appears
to be of paramount importance for the maintenance of symbioses among
marine organisms of diverse phyla. Ross (see page 118) reviews the
importance of contact chemical stimuli in maintaining anemone-hermit
crab relationships. Chemical "shell factor" also serves to mediate
"contact" recognition of the molluscan host, *Tegula funebralis* by
the limpet, *Acmaea asmi* (Alleman 1968). Most experimental investigations, however, focus on "distant" host recognition using some
modification of the Y-maze or trough-type olfactometers (Varley and
Edwards 1953). A diffusible substance from the horseshoe crab,
Limulus, attracts the symbiotic triclad, *Bdelloura caudida*, in a
Y-maze assay system.* Polychaetes associated with echinoderms[†]
(Dimock and Davenport 1971) and molluscs (Gerber and Stout 1968;
Webster 1968) select the arm of Y-type mazes containing "host factor"
over one containing plain seawater. Crustaceans associated with
molluscs (Webster 1968; Ache and Davenport 1972) and polychaetes
(Carton 1968) respond in a similar manner to effluents of their
respective hosts. Crustaceans associated with echinoderms orient
towards "host factor" in a trough olfactometer (Gray, McCloskey, and
Weihe 1968), an apparatus also used to demonstrate chemically mediated host recognition in pelecypods associated with echinoderms,
polychaetes, and sipunculids (Gage 1968).
   Understanding the role of chemically mediated information in
host location requires a knowledge of the extent to which such information can account for the restricted host specificity that charac-
terizes many marine symbioses. In some associations, chemical stimuli contain sufficient information to account for the naturally occurring specificity of the relationship. Four populations of the polynoid worm, *Arctonoe pulchra*, each associated with a different host

---

* David Boylan: personal communication.
† William Stewart: personal communication.

species, preferentially respond to effluents of their respective hosts in a Y-maze olfactometer, when tested against both plain seawater and effluents of other organisms. Worms from the seastar, *Dermasterias imbricata*, two species of holothuriods, *Stichopus* spp., and the gastropod, *Megathura crenulata*, failed to respond to effluents from test organisms closely related to their respective original hosts or to effluents from species which function as alternate hosts, species which in turn are attractive as hosts for their own respective populations of symbiotic polynoids (Dimock and Davenport 1971). Gray, McCloskey, and Weihe (1968) also used a trough olfactometer to demonstrate extreme specificity in the ability of the crab, *Dissodactylus mellitae*, associated with the echinoid, *Mellita* sp., to chemically discriminate its host from closely related echinoderms. Of six species of echinoderms, only another echinoid, *Encope* sp., elicited a response, but one significantly weaker than that elicited by the natural host. Both field and laboratory studies, however, suggest that specimens of *Encope* may alternatively host the crab. Another crustacean, the copepod, *Sabelliphilus sarsi*, appears capable of effecting extreme host specificity by chemically discriminating its natural host, the polychaete, *Spirographis spallanzani*, from two nonhost but congeneric polychaetes. *S. pavonia* elicits no attraction while the conspecific *S. spallanzani* var. *brevispira* elicits only a weak attraction relative to that of the natural host (Carton 1968). The results of these three studies support the earlier reports that chemical cues of host origin can be highly specific in their action.

The findings of Dimock and Davenport (1971) on *Arctonoe* symbioses, however, add a new dimension to the concept of chemically mediated host specificity. These investigators demonstrate that response specificity to chemical attractants can be modified by the previous experience of the symbionts. Worms could be conditioned by long-term (2-4 weeks) physical contact with an alternate host to respond preferentially to effluents from this organism, effluents previously unattractive or only weakly stimulating to the worms. Interestingly, choice preference could not be altered by prolonged olfactory exposure; physical contact between the symbiont and the alternate host was necessary during the conditioning period. Gray, McCloskey, and Weihe (1968) report a functionally similar "acclimation" of the symbiotic crab, *Dissodactylus*, toward the otherwise weakly attractive effluent of the alternate host echinoid, *Encope* sp. The role of chemically mediated information in maintaining a symbiosis may thus be different from its role in initiating the same relationship. This possibility should be rigorously quantitied in future studies of chemically mediated host specificity.

The author's studies with the symbiotic caridean, *Betaeus macginitieae*, suggest that constraints in addition to those imposed by chemical stimuli are operating to effect the observed specificity of its association with the sea urchins, *Strongylocentrotus* spp. (Ache and Davenport 1972). Shrimp behavior was quantified in a two-celled choice apparatus designed to be compatible with the fast-moving *Betaeus* yet retain the binomial simplicity of a conventional Y-maze. The apparatus (Fig. 1) incorporated a 33 x 46 x 12 cm opaque white polyethylene pan fitted with a T-shaped transparent plastic divider to form two small compartments (referred to as test or control cells) and a larger compartment (referred to as the choice

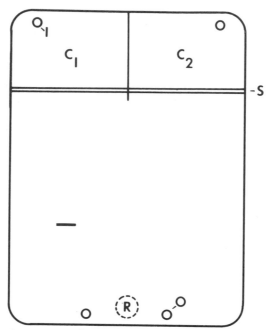

Figure 1. Diagram (top view) of two-celled choice apparatus: C_1 and C_2, test or control cells; I, seawater inlets; O, seawater outlets; S, transparent plexiglass screen, with opening along lower edge to allow water movement into the choice area; R, removable release cylinder. Bar represents approximate adult size of the shrimps tested in this apparatus.

area). A baffled opening in the transverse partition allowed seawater introduced into the test and control cells to flow into the choice area where it was removed by two constant-level siphons. An input of 7.0 ml/sec of new seawater to each cell produced an even, laminar flow of approximately 5 mm/sec along the bottom of the choice area. A removable opaque cylinder allowed introduction of single shrimp into the choice area with minimal directional bias. Thirty shrimps, selected at random from groups of seventy to eighty individuals, were utilized in each experiment. Table 2 summarizes the results of experiments extending over five consecutive days that permitted specimens of *B. macginitieae* collected from the urchin, *S. franciscanus*, to discriminate between a test cell containing individuals of one of seven different species of test organisms and a control cell containing only seawater. An opaque white plastic screen placed over the transverse partition occluded visual stimuli

TABLE 2
Specificity of chemically-mediated host location: *B. macginitieae*

Experiment No.	Contents of test cell	Shrimp choosing Test	Shrimp choosing Control	χ^2 Choice	P
1	*S. franciscanus*	25	0	25.0	<0.005
2	Seawater	4	2	0.66	0.25-0.50
3	*S. purpuratus*	20	0	20.0	<0.005
4	*L. anamesus*	5	1	2.66	0.10-0.25
5	*S. parvimensis*	8	6	0.57	0.25-0.50
6	*D. imbricata*	8	6	0.57	0.25-0.50
7	*H. rufescens*	9	6	0.60	0.25-0.50
8	*U. caupo*	8	5	0.69	0.25-0.50
9	*S. franciscanus*	21	1	18.2	<0.005
10	Seawater	3	5	0.50	0.50-0.75

from the choice situation without altering the flow characteristics
of the apparatus. The mean weight of the test organisms utilized
in these experiments was 200±25 g. Only effluent of the congeneric
echinoids, *Strongylocentrotus franciscanus*, the natural host
(Experiments 1, 9), and *S. purpuratus* (Experiment 3), elicited differential choice between test and control cells. Differential
choice was not elicited by effluents of the nonhost echinoid, *Lytechi-
nus anamesus* (Experiment 4), nor the nonechinoid echinoderms, *Stichopus parvimensis* (Experiment 5) and *Dermasterias imbricata* (Experiment
6). Likewise, effluents of the abalone, *Haliotis rufescens* (Experiment 7), and the echiurid worm, *Urechis caupo* (Experiment 8), both
reported to be hosts for other *Betaeus* species (Hart 1964), failed
to elicit differential choice. A test of association indicates that
the final distribution elicited by effluents of the urchin, *S. purpuratus* (Experiment 3), does not differ significantly from the more
extreme of the two distributions elicited by effluents of the natural
host (Experiment 1) (χ^2 - 1.46, P - 0.1 - 0.25). Agreement of initial and final repetitions of the basic host response (Experiments
1, 9) supports the hypothesis that no change occurred in the response
level of the shrimps during the duration of the experimental period.

    This set of experiments suggests that chemically mediated infor-
mation is sufficient to account for genus-specific host recognition
by *B. macginitieae*. These shrimps, however, associate predominately
with the homochromous giant red sea urchin, *S. franciscanus*, and only
secondarily with the smaller purple congeneric, *S. purpuratus* (Ache
1970). It is necessary to hypothesize that some additional constraint is active in the resolution between the primary and secondary
urchin hosts. Visual stimuli also mediate host location in this
relationship (Ache and Davenport 1972) and provide sufficient
information to explain the shrimps' ability to discriminate between
the two host urchins. These experiments are described under the
heading "Visually mediated host location".

Understanding the role of chemically mediated information in host location also requires knowledge of the ability of chemical stimuli to communicate the directional information required for host location. A question arises as to the ability of chemical stimuli alone to effect directed locomotion toward an odor source in the marine environment (Fraenkel and Gunn 1961; Gage 1966a). In the marine benthos particularly, local water turbulence and surging would disrupt diffusion gradients required for chemotaxic orientation. This question has not been critically analyzed, as both Y-maze and trough olfactometers do not necessarily present the symbiont with a concentration gradient of the attractant. One exception is the short, very steep gradient in the trough olfactometer at the edge of an attractant-containing stream. In both these devices, the chemical attractant is delivered in a carrier current which allows the alternative explanation that "upstream" orientation is the result of the symbiont's exhibiting a positive rheotaxis in in the presence of an appropriate chemical cue. The role of rheotaxis in host location is discussed under the heading, "Mechanically mediated host location".

VISUALLY MEDIATED HOST LOCATION

Visual stimuli have seldom been investigated as possible active components in the host location behavior of invertebrate symbionts, although several fish partnerships are maintained, at least in part, by visually mediated information (Magnus 1967; Dix 1969; Trott 1969; Losey 1971). Davenport (1966) noted that the active movements and bright color patterns of many cleaning shrimps strongly suggest visual stimuli are active in maintaining these associations but, to the author's knowledge, this hypothesis has yet to be tested experimentally.

Experiments by the author on the symbiotic caridean, *Betaeus macginitieae*, indicate that these shrimps can effectively locate their host urchins using either visually or chemically mediated information of host origin (Ache and Davenport 1972). Table 3 summarizes the results of experiments demonstrating this point; the

TABLE 3
Role of chemical and visual stimuli in host location: *B. macginitieae*

Experiment No.	Contents of test cell	Stimulus modalities present[*]	Shrimp choosing		χ^2 Choice	P
			Test	Control		
1	*S. franciscanus*	V,C	28	1	25.0	< 0.005
2	*S. franciscanus*	V	22	0	22.0	< 0.005
3	*S. franciscanus*	C	22	1	19.2	< 0.005
4	Seawater	-	2	2	--	--

[*]V=visual; C=chemical; - = neither visual or chemical.

two-celled choice apparatus (Fig. 1), which permitted speciments of
B. macginitieae collected from the urchin, *S. franciscanus*, to
select between a test cell containing the host urchin and a control
cell containing seawater only, was used. In the basic stimulus
situation, where both chemical and visual stimuli were experimentally
unaltered, significantly more shrimp selected the host-containing
cell (Experiment 1). With the host urchin contained in a clear
glass 4-1 beaker placed in the test cell and a seawater-filled beaker
placed in the control cell, significantly more shrimps still selected
the host-containing cell (Experiment 2). With an opaque white plastic screen attached to the transverse partition, thus masking visual
stimuli from the choice situation while not interfering with chemical
stimuli, significantly more shrimps again selected the host-containing
cell (Experiment 3). Neither stimulus modality acting alone, however,
elicited host location to the extent that both did when presented
together. Very few shrimps made a choice in the absence of any
stimuli of host origin (Experiment 4).

Further experiments indicate that the visually mediated information effecting host location is not completely sufficient to
account for the discrimination of the host urchin, *S. franciscanus*,
from other test organisms (Ache 1970). These experiments utilized
a modification of the two-celled choice apparatus (Fig. 1) in which
the open-bottomed transparent partition was replaced with a water-
tight transparent partition to insure chemical isolation of all three
compartments and the seawater inlet tubes were removed to create a
static system. Table 4 summarizes the results of these experiments
in which specimens of *B. macginitieae* were tested for their inability
to discriminate between a cell containing the host urchin and a cell
containing one of four other test organisms, all organisms being
equated by displacment volume (± 50 cc) to that of the single *S.
franciscanus* used in the pairings. Significantly more shrimps
selected *S. franciscanus* over the congeneric urchin, *S. purpuratus*
(Experiment 4), as well as the gastropod, *Haliotis rufescens* (Experiment 1), and the holothuroid, *Stichopus parvimensis* (Experiment 2).
Interestingly, shrimps did not discriminate between the black giant
keyhole limpet, *Megathura crenulata*, and the host urchin (Experiment
3). This latter observation suggested the possibility that the
urchin shrimps are responding to a dark solid form of unspecific

TABLE 4
Specificity of visually-mediated host location: *B. macginitieae*

Experiment No.	Contents of Cell 1	Shrimp choosing		Contents of Cell 2	P
		Cell 1	Cell 2		
1	*S. franciscanus*	28	2	*H. rufescens*	<0.05
2	*S. franciscanus*	27	3	*S. parvimensis*	<0.05
3	*S. franciscanus*	17	13	*M. crenulata*	>0.05
4	*S. franciscanus*	26	4	*S. purpuratus*	<0.05
5	*S. franciscanus*	17	13	*S. franciscanus*	>0.05

peripheral outline and not to any visually mediated information uniquely characterizing the host urchin, *S. franciscanus*. This hypothesis was confirmed in a series of experiments requiring shrimps to discriminate among two-dimensional models in which parameters could be more rigorously controlled (Ache and Davenport 1972).

Visually mediated information can account for the urchin shrimps' discrimination between the two congeneric urchin hosts, *S. franciscanus* and *S. purpuratus*, a degree of discrimination not possible via chemically mediated information of host origin. Thus, the natural specificity of the *Beteaus-Strongylocentrotus* appears to be the result of bimodal information transfer from the host to the symbiont. Neither stimulus modality, acting alone, can adequately account for the restricted specificity of this association. As suggested by the results of the chemical/visual discrimination experiments (Table III, Experiments 2, 3 versus Experiment 1), this bimodal information may act synergistically to effect host location. The mechanism by which information contained in these two channels is utilized by the animals, however, remains to be determined.

MECHANICALLY MEDIATED INFORMATION

Mechanically mediated information is potentially active in controlling host recognition once contact is established between potential partners, although this hypothesis has not been subject to experimental analysis. Castro (1969) noted that adult crabs, *Echinoecus pentagonus*, symbiotic with the echinoid, *Echinothrix calamaris*, are repelled by the vigorous spine movements elicited from a nonhost diadematid urchin, *Diadema paucispinum*, when the two organisms are placed in contact with each other. Such recognition of "nonhost" may be potentially important in the maintenance of host specificity in other relationships and should be subject to further investigation. "Rugophilic" behavior, the preference of organisms for confined spaces such as grooves and crevices, is known to be an active component in the settlement of larvae of several nonsymbiotic marine organisms (Wisely 1969).

The idea that mechanically mediated information may also be active in "distant" host recognition was enhanced by the interesting findings of Gilpin-Brown (1969) that searching movements in settled juvenile *Nereis fucata* can be initiated by low frequency substrate-borne vibrations. These movements are identical to those elicited by the shell-dragging locomotion of host hermit crabs and presumably represent an active component of the worms' host location response. The author is not aware of any other studies, however, that demonstrate an active role of mechanical stimuli of host origin effecting host location.

As noted in the previous section on chemically mediated host location, information transmitted by water currents carrying chemical stimuli of host origin may be active in communicating the direction of the odor source (the host) to the symbiont. Gray, McCloskey, and Weihe (1968) demonstrated that the crab, *Dissodacty-lus*, will select an arm of a Y-maze containing a stronger flow of plain seawater over one containing "host factor." The palaemonid shrimp, *Anchistus custos*, symbiotic with the bivalve *Pinna*, was only observed to abandon random movement and swim directly and rapidly to its host when sudden movements of the *Pinna* produced a jet of water which impinged on the shrimp. This behavior could be duplicated by jets of water produced by pipettes in the presence of a host (Johnson and Liang 1966). Although Davenport, Camougis, and Hickok (1960) were not able to demonstrate chemically induced positive rheotaxis in either a symbiotic polychaete or a pinnotherid crab, Gage (1966a, 1968) used a similar trough-type olfactometer with several species of erycinacean bivalves and demonstrated such a response to water currents containing a chemical stimulus of host origin.

Studies of a similar nature by the author suggest that the caridean, *Betaeus harfordi*, locates its gastropod hosts, *Haliotis* spp., by responding to water currents, in the presence of appropriate chemical cues of host origin (Ache and Davenport 1972). These shrimps, unlike the congeneric urchin symbionts (*B. macginitieae*) previously discussed, do not effect host location with visually mediated information of host origin. Table 5 (Series 1) summarizes the results of experiments using the two-celled choice apparatus (Fig. 1) that demonstrate this point. Shrimps collected from the host gastropod, *H. corrugata*, preferentially selected the host-containing cell only in the presence of chemical stimuli of host origin (Experiments 1, 2). With chemical stimuli removed from the

TABLE 5
Role of chemical, visual and water current stimuli in host location: *B. harfordi*

Experiment No.	Contents of test cell	Stimulus modalities present[*]	Shrimp choosing		χ^2 Choice	P
			Test	Control		
Series 1						
1	*H. corrugata*	V,DC	26	1	23.0	<0.005
2	*H. corrugata*	DC	25	2	19.6	<0.005
3	*H. corrugata*	V	14	6	1.60	0.10-0.25
4	Seawater	--	4	4	--	--
Series 2						
1	*H. rufescens*	V,DC,C	26	0	26.0	<0.005
2	*H. rufescens*	V,C	2	2	--	--
3	*H. rufescens*	V	4	3	0.14	0.50-0.75
4	*H. rufescens*	V,NDC	8	0	8.00	<0.005
5	*H. rufescens*	V,NDC,C	14	10	0.67	0.25-0.50

[*]V=visual; DC=directed chemical; NDC=non-directed chemical; C=current.

choice situation by containing the abalone in a clear glass 4-1 beaker and testing it against an identical seawater-filled beaker placed in the control cell, preferential selection of the host-containing cell when only chemical stimuli were present (Experiment 2) as when both visual and chemical stimuli were present, i.e., the experimentally unaltered situation (Experiment 1).

To gain a fuller understanding of the results of Series 1 experiments, it was necessary to know if the presence of chemical stimuli triggered a response to current, since chemical stimuli were always presented in association with a directional flow of water emanating from the test and control cells. Table 5 (Series 2) summarizes the results of experiments designed to answer this question with shrimps collected from the host gastropod, *H. rufescens*. Experiment 1 is a replicate of the experimentally unaltered choice situation in Series 1. To determine if current alone had any effect, shrimps were permitted to discriminate between a test cell containing a model abalone and a seawater control cell both with (Experiment 2) and without (Experiment 3) a current in the apparatus. To eliminate current, the seawater inlets were closed. Substituting a model abalone for a live one in the test cell allowed presentation of visual stimuli without chemical stimuli while retaining the directional flow. The model consisted of a paraffin-filled abalone shell with 1.5 cm wide "epipodium" of black tape exposed beneath the ventral edge of the shell. As live abalone remained stationary when placed in the apparatus, a static model was judged an acceptable substitute. As can be seen, differential choice between test and control cells was not elicited in either experiment. It appears that current itself does not affect the activity of shrimps or their response to visual stimuli. To determine if the presence of non-directional chemical stimuli had any effect, shrimps were permitted to discriminate between test and control cells when nondirectional chemical stimuli of host origin were present throughout the system, but in the absence of a flow (Experiment 4). Two specimens of *H. rufescens* (350 g), confined in a perforated plastic cup and swirled in the choice area of the apparatus for 1 min prior to introduction of each shrimp, served to introduce nondirectional chemical stimuli into the system. Assuming the attractant had a time stability of at least 6.0 min, it was present in the choice area throughout the time interval allowed for choice. This assumption, of course, could only be confirmed by a positive result, i.e., by obtaining a significant change in response on the addition of the attractant. Differential choice was elicited in favor of the model-containing cell, suggesting that the presence of nondirectional chemical stimuli may enhance the stimulus value of visual cues characterizing the model host. A test of association comparing the number of shrimps choosing the model-containing cell versus the number not choosing it in this (Experiment 4) and in the control situation with no current and no chemical (Experiment 3), however, indicates that no significant increase in the level of activity occurred in the presence of the nondirectional chemical stimuli (χ^2 - 0.178, P - 0.50 - 0.75). It appears that nondirectional chemical stimuli alone are not sufficient to affect the activity of the shrimps. To determine if nondirectional chemical stimuli serve to trigger a response to current, shrimps were permitted to discriminate between test and control cells when nondirectional chemical stimuli of host origin were presented simultaneously with a directional flow (Experiment 5). No differential choice was elicited between the test and control cells. However, the method of introducing the chemical stimuli in this experiment should have dispersed host effluent throughout all compartments of the apparatus. Since a current was flowing under the transparent partition from both test and control cells, no difference should have existed for the visual stimuli of the model-containing cell. Experiments 2 and 3 indicate that visual stimuli with or without current elicit little activity. Thus, the combined number of shrimps reaching either cell can be considered as being most characteristic of the response to this stimulus situation. A test of association comparing the total number of shrimps choosing either cell versus the number not choosing either cell in Experiment 5 and in the basic host response (Experiment 1 - visual, current, and directed chemical stimuli of host origin) indicates no significant difference in the level of activity (P>0.995). Host-oriented locomotion in the shrimp *B. harfordi* appears to result from the ability of chemical stimuli to release an oriented response to directional water currents.

DISCUSSION

Chemically mediated information appears to dominate the sensory input controlling host location behavior among "lower" symbiotic marine invertebrates. Among "higher" invertebrates, the possibility must be considered that stimuli of other modalities complement (perhaps "replace" in some associations, e.g. the *Nereis-Eupagurus* association) the role of chemically mediated information in controlling host location. A multimodal sensory basis of host location would maximize the information available to the symbiont while it would not demand excessive competence in its sensory system for any one modality. Such a mechanism of information transfer agrees well with the general concept of parsimonious invertebrate systems. Thus, species-specific rather than genus-specific host discrimination is possible in the *Betaeus-Strongylocentrotus* association via the combined information of visual and chemical stimuli. A relatively generalized visual response reduces the discriminatory ability required of the distance chemoreceptors. Similarly, chemoreceptive organs mediating host-oriented locomotion in the *Betaeus-Haliotis* association must discriminate only the presence or absence of the attractant and not small changes in stimulus intensity (concentration) otherwise required to effect a chemotaxic mechanism of locomotion.

Our understanding of the mechanisms by which symbiotic partners come together is still limited, however, in spite of the experimental advantages symbioses offer to students of invertebrate behavior. Several questions arise as logical extensions of the recent experimental analyses of marine symbioses: (1) What is the chemical nature of the attractants? The highly specific nature of many symbiotic relationships should provide an excellent opportunity to investigate information transfer via chemical stimuli. To this end, William Stewart* has partially analyzed the attractant of the *Podarke* (Ophiodromus)-*Patiria* association and found two active components, each of molecular weight less than 500, one eliciting a generalized increase in the worms activity, the second eliciting a "chemotaxis" when presented to the worms in a carrier current of seawater. (2) By what mechanism(s) do multimodal stimuli control host location behavior? The effects of multimodal information could result from sequential action of the individual modalities or from simultaneous summation of several modalities. Behavioral analyses of this question are necessary prerequisites to further electrophysiological studies of host-oriented behavior. (3) How plastic is the specificity of these associations? This question, of course, requires successful rearing of the symbiont through all stages of its life cycle, a difficult task with many types of invertebrates, but one that should benefit from the recent emphasis of aquaculture of marine organisms. This question is particularly interesting in light of the conditioning experiments of Dimock and Davenport previously mentioned. Hopefully, future studies will provide answers to these and related questions on host-symbiont interactions.

As noted by Marler and Hamilton (1966),"Analysis of the stimulus properties responsible for the environmental control of different action patterns is a central concern in the study of animal behavior." Knowledge of the stimuli initiating and maintaining symbiotic relationships, coupled with advances in the closely related studies of food-finding, habitat selection, mate recognition, and prey-predator interactions will provide a sensory basis for understanding behavior in marine organisms.

---

*William Stewart: personal communication.

LITERATURE CITED

Ache, B. 1970. An analysis of the sensory basis of host recognition in symbiotic shrimps of the genus *Betaeus*. Ph. D. thesis, Univ. of California.

_____; and Davenport, D. 1972. The sensory basis of host recognition by symbiotic shrimps, genus *Betaeus*. Biol Bull. 143:94-111.

Alleman, L. 1968. Factors affecting the attraction of *Acmaea asmi* to *Tegula funebralis*. Veliger 11 (Suppl):61-63.

Carton, Y. 1968. Specificite parasitaire de *Sabelliphilis sarsi*, parasite de Spirographis spallanzani. III. Mise en evidence d'une attraction biochemique du copopode par l'Annelide. Arch. Zool. Exp. et Gen. 109:123-44.

Castro, P. 1969. Symbiosis between *Echinoecus pentagonus* (Crustacea, Brachyura) and its host in Hawaii, *Echinothrix calamaris* (Echinoidea). Ph. D. dissertation, Univ. of Hawaii.

Cheng, T. 1967. Marine molluscs as hosts for symbioses. In Advances in marine biology, ed. F. Russell, Vol. 5. New York: Academic Press.

Davenport, D. 1955. Specificity and behavior in symbioses. Quart. Rev. Biol. 30:29–46.

_____. 1966. The experimental analysis of behavior in symbioses, pp. 381–429. New York: Academic Press.

_____, Camougis, G., and Hickok, J. 1960. Analyses of the behavior of commensals in host-factor. 1. A hesionid polychaete and pinnotherid crab. Anim. Behav. 8:209–18.

Dimock, R., Jr., and Davenport, D. 1971. Behavioral specificity and the induction of host recognition in a symbiotic polychaete. Biol. Bull. 141:472–82.

Dix, T. 1969. Association between the echinoid *Evechinus chloroticus* (Val.) and the clingfish *Dellichthys morelandi* Briggs. Pac. Sci. 23:332–35.

Fraenkel, G., and Gunn, D. 1961. The orientation of animals, 2nd ed. New York: Dover Publ.

Gage, J. 1966a. Experiments with the behaviour of the bivalves, *Montacuta substriata* and *M. ferruginosa*, 'commensals' with spatangoids. J. mar. biol. Ass. U. K. 46:71–88.

_____. 1966b. The life histories of the bivalves *Montacuta substriata* and *M. ferruginosa*, 'commensals' with spatangoids. J. mar. biol. Ass. U. K.:499–511.

_____. 1968. The mode of life of *Mysella cuneata*, a bivalve 'commensal' with *Phascolion strombi* (Sipunculoidea). Can. J. Zool. 46:919–34.

Gerber, H., and Stout, J. 1968. Sensory basis of the symbiotic relationship of *Arctone vitta*(Grube)(Polychaeta, Polynoidae) to the keyhole limpet, *Diadora aspera*. Physiol. Zool. 41:169–79.

Gilpin-Brown, J. 1969. Host-adoption in the commensal polychaete, *Nereis fucata*. J. mar. biol. Ass. U. K. 49:121–27.

Gray, I., McCloskey, L., and Weihe, S. 1968. The commensal crab, *Dissodactylus mellitae* and its reaction to sand dollar host-factor. J. Elisha Mitchell Sci. Soc. 84:472–81.

Harris, L. 1971. Nudibranch associations as symbioses. In Aspects of the biology of symbiosis, ed. T. Cheng, pp. 77–90. Baltimore: Univ. Park Press.

Hart, J. 1964. Shrimps of the genus *Betaeus* on the Pacific coast of North America with descriptions of three new species. Proc. U. S. Nat. Mus. 115:431–66.

Jensen, K. 1970. The interaction between *Pagurus bernhardus* (L.) and *Hydractinia echinata* (Fleming). Ophelia 8:135–44.

Johnson, B., and Liang, M. 1966. On the biology of the watchman prawn, *Anchistus custos* (Crustacea; Decapoda; Palaemonidae),and Indo-West Pacific commensal of the bivalve *Pinna*. J. Zool. London 150:433–55.

Liang, J. 1937. Host-finding by insect parasites. I. Observations on the finding of hosts by *Alysia manducator, Mormoniella vitripennis* and *Trichogramma evancescens*. J. Anim. Ecol. 6: 298-317.

Losey, G., Jr. 1971. Communication between fishes in cleaning symbiosis. In Aspects of the biology of symbiosis, ed. T. Cheng, pp. 45-76. Baltimore: Univ. Park Press.

McCauley, J. 1969. Marine invertebrates, chemical signals, and marine products. Lloydia 32:425-37.

Magnus, D. 1967. Ecological and ethological studies and experiments on the echinoderms of the Red Sea. Studies Trop. Oceanogr. 5: 635-64.

Marler, P., and Hamilton, W. 1966. Mechanisms of animal behavior. New York: J. Wiley & Sons.

Pijl, van der, L., and Dodson, C. 1966. Orchid flowers, their pollination and evolution. Coral Gables: Univ. Miami Press.

Trott, L. 1969. Contributions to the biology of carapid fishes (Paracanthopterygii: Godiformes). Univ. Calif. Publ. Zool. 88: 1-39.

Ulmer, M. 1970. Site-finding behavior in helminths in intermediate and definitive hosts. In Ecology and physiology of parasites, ed. A. Fallis, pp. 123-60. Toronto: Univ. Toronto Press.

Varley, G., and Edwards, R. 1953. An olfactometer for observing the behaviour of small animals. Nature 171:789-90.

Webster, S. 1968. An investigation of the commensals of *Cryptochiton stelleri* (Middendorff, 1946) in the Monterey Peninsula area, California. Veliger 11:121-25.

Wisely, B. 1969. Preferential settlement in concavities (rugophilic behavior) by larvae of the brachiopod, *Waltonia inconspicua* (Sowerby, 1946). Aust. J. mar. Freshwat. Res. 3:273-80.

Wright, C. 1959. Host location by trematode miracidia. Ann. Trop. Med. Parasitol. 53:288-92.

Cellular reactions in marine pelecypods as a factor influencing endosymbioses

Thomas C. Cheng, Ann Cali, and David A. Foley
Institute for Pathobiology and Department of Biology
Lehigh University
Bethlehem, Pennsylvania

Endosymbiosis in marine pelecypods, or in all categories of hosts for that matter, may be topographically categorized into four general types: intracellular, intercellular in tissues, intraluminal within some segment of the alimentary tract and its associated organs and ducts, and within other body cavities. Of course, in the case of acoelomate invertebrates the last-mentioned habitat is unavailable, and in certain groups of hosts, such as the protozoans, poriferans, and cnidarians, an alimentary tract, in the strict sense of the term, is also not available as a site for symbiosis. As Read (1971) has pointed out, the physico-chemical nature of the vertebrate's alimentary tract is of critical importance for the successful establishment of endosymbiosis at this site and the characteristics of the various components of the alimentary tract are governed to a large extent by the host's physiologic and metabolic activities. The same holds true for invertebrate hosts, although both qualitative and quantitative measurements of their milieu interieur have in most instances not yet attained the degree of exactness as in the case of vertebrates, especially mammals. Nevertheless, the principle is recognized that a thorough understanding of any symbiotic relationship must involve investigations of both the symbiont and its habitat (i.e., its host).

In addition to the physico-chemical characteristics of the habitat, most symbiologists, especially parasitologists, now realize that another dynamic feature of the host, its ability to recognize "self" from "nonself," is also of critical importance in influencing the establishment or nonestablishment of endosymbiosis. The intent of this paper is to review some of the more salient features of one aspect of this concept as portrayed by marine molluscs, that of phagocytosis and related processes, and to contribute some new related information.

CELLULAR REACTIONS IN MOLLUSCS

The various types of cellular reactions in molluscs to foreign agents naturally or experimentally introduced into their bodies have been reviewed in recent years by Stauber (1961), Cheng and Sanders (1962), Cheng (1967), Feng (1967), and Cheng and Rifkin (1970). In brief, particulate and soluble materials recognized as being foreign are usually phagocytized unless their physical sizes prevent their being engulfed by the host's leucocytes, in which case they become encapsulated. Several types of encapsulation have now been recognized on a structural basis (Cheng and Rifkin 1970)(Table 1),although the physiological processes leading to each remain to be elucidated.

TABLE 1
Known types of encapsulation occurring in molluscs as the result of invasion by metazoan parasites[*]

---

1. *Antiqufibrous encapsulation.* Involving fibrous elements which are not formed *de novo* as the result of parasitic stimulation but represent preexisting fibers present in the immediate vicinity of the parasite.

2. *Novufibrous encapsulation.* Involving fibrous elements which are formed *de novo* as the result of parasitic stimulation.

3. *Fibroblastic encapsulation.* Capsule formed of fibroblasts or fibroblast-like cells. No true fibers are involved.

4. *Leucocytic encapsulation.* Involving the aggregation of leucocytes to form a tunic surrounding the parasite.

5. *Myofibrous encapsulation.* Capsule formed from preexisting muscle cells present in the immediate vicinity of the parasite.

---

[*]For a detailed discussion, see Cheng and Rifkin, 1970.

In certain special cases, nacrezation may be the type of cellular reaction evoked. This process involves the secretion of nacre by the host's mantle around the foreign agent (see reviews by Tsujii 1960, and Cheng and Rifkin 1970). The major criterion for the induction of nacrezation appears to be physical contact with the nacre-secreting areas of the mantle. For example, it is known that certain metacercariae, such as that of *Meiogymnophallus minutus (=Gymnophallus margaritarum)* occurring between the inner surface of the shell and the mantle of various marine pelecypods will cause this type of reaction (Dubois 1901, 1907; Perrier 1903; Jameson 1902; Giard 1907; and others).

Although the fate of phagocytized, encapsulated, or nacrezized organisms is commonly death through resorption, this is not consistently the case, especially in reference to phagocytosis. It is known, for example, that certain bacteria are not only sustained within host leucocytes for considerable time but some actually multiply intracellularly (Michelson 1961). In addition, there are reports that the fungus *Labyrinthomyxa marina* can survive as an intracellular parasite (Prytherch 1940; Mackin 1951). Furthermore, it has been assumed that certain species of zooxanthellae can become established as intracellular mutualists in their molluscan hosts. This is most conspicuously demonstrated by zooxanthellae in cells of the marine tridacnid molluscs *Hippopus* and *Tridacna* (Yonge 1936, 1953). It must be emphasized that Yonge, as a result of his histological studies, is of the opinion that the intracellular habitat of the dinoflagellates within the host's phagocytes is only temporary, i.e., being limited to that period during which they are transported from the siphonal hemal sinuses where they are "farmed" to the interacinar spaces of the digestive gland.

Not all endosymbiotic algae, however, are intracellular. For example, Goetsch and Scheuring (1926) and Buchner (1965) have reported that the zoochlorellae in such freshwater molluscs as *Lymnaea*, *Anodonta*, and *Unio* are seldom within host cells. They are extracellular in tissues.

The available information pertaining to the occurrence of intracellular symbionts in molluscan phagocytes raises some interesting questions. Among these are: (1) Are molluscan phagocytes attracted to foreign materials, including potential symbionts, and if so, by what? (2) What factors permit the survival and propagation of intracellular symbionts within molluscs? (3) How are incompatible invading organisms degraded intracellularly? (4) Is the intracellular degradation of incompatible symbionts of any value to the host other than the obvious elimination of foreign invaders? Some speculations, interpretations, and partial answers follow.

ATTRACTION OF MOLLUSCAN LEUCOCYTES TO FOREIGN MATERIALS

It is well documented that molluscan leucocytes will not only increase in number (leucocytosis) but also confront foreign substances that have been experimentally introduced (Stauber 1950; Tripp 1958a, b, 1960, 1961; Feng 1959, 1965, 1966, 1967; Cheng 1966, 1967; Cheng and Cooperman 1964; Arcadi 1968; Cheng and Rifkin 1970; Cheng, Shuster, and Anderson 1966a; Cheng, Thakur, and Rifkin 1969; Pauley and Drassner 1972). Although a number of investigators have suggested that the host's leucocytes may be attracted to the foreign substances, biotic and abiotic, this has remained a moot point, although Cheng, Shuster, and Anderson (1966b) have reported as the result of in vitro studies that it is the newly formed metacercarial cyst of *Himasthla quissetensis* that attracts the leucocytes of the American oyster, *Crassostrea virginica* (Fig. 1), as well as the cells of the other marine pelecypods tested.

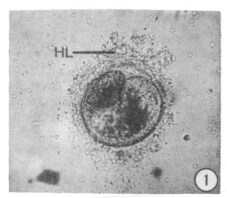

Figure 1. Recently encysted *Himas-thla quissetensis* meta-cercaria in *Crassostrea virginica* plasma to which leucocytes had been added. Notice the large number of leucocytes (HL) adher-ing to the surface of the metacercarial cyst (x 20 obj., x 10 ocular).(After Cheng, Shuster, and Anderson 1966b; with per-mission of Academic Press.)

In order to add further credence to the concept that chemotaxis plays a role in attracting host cells to foreign agents, the follow-ing study was carried out.

Materials and Methods

Since it is known that the leucocytes of several species of estuarine pelecypods will form a cellular reaction tunic (encapsula-tion) around encysted *Himasthla quissetensis* metacercariae in vivo (Cheng et al. 1966a) and that the leucocytes of *Crassostrea virginica* will adhere to the outer surface of *H. quissetensis* metacercarial cysts in vitro (Cheng et al. 1966b), a test chamber was designed that would permit the quantification of host cells migrating towards meta-cercarial cyst walls. This apparatus*, constructed out of plexiglass, consists of an upper and a lower cylindrical chamber (Fig. 2). The lower chamber measures 8 mm (inside diameter) by 2 mm while the upper one measures 8 mm (inside diameter) by 10 mm. A 1.2 µ Millipore fil-ter, prewashed in 70% ethanol and thoroughly rinsed in distilled water, is placed between the two chambers while freshly collected *H. quissetensis* metacercarial cyst fragments from 300 metacercariae were placed in the upper chamber suspended in 20 o/oo filtered seawater. Approximately 550 leucocytes, freshly collected from the adductor muscle sinus of *C. virginica* with syringe and hypodermic needle, were placed in the lower chamber suspended in sufficient homologous serum to completely fill it. After tightly sealing the system, the appara-tus was permitted to stand for 3 hours at 22°C after which the Milli-pore filter was removed, stained with hematoxylin and eosin, cleared in xylene, and mounted on glass slides for microscopical examination. The number of leucocytes attached to the filter was ascertained in

---

*This apparatus was designed and constructed by George P. Hoskin of this Institute.

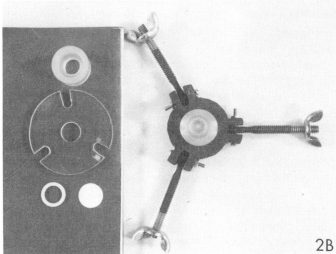

Figure 2.
Apparatus employed to test the chemotactic response of *Crassostrea virginica* leucocytes to *Himasthla quissetensis* metacercarial cysts. A, assembled apparatus; U, upper chamber; L, lower chamber; B, apparatus disassembled to show component parts.

each instance. As controls, an identical number of oyster leuco-cytes was placed in the lower chamber, but the upper chamber was filled with seawater.

Results and Comments

The results of one series of our experiments are presented in Table 2. It is noted, however, that although the presented data indicate a significant difference between the experimental and con-trol groups, thus indicating the attraction of leucocytes to the cyst material, on another occasion the results obtained were not as drama-tic. Although we are convinced that there is an attraction of oyster leucocytes to the metacercarial cyst material, it must be mentioned that the results are erratic. The reason for this is presented at a later point. It remains to be determined what is the chemical con-stituent(s) of cyst walls that attracts the mollusc's leucocytes.

TABLE 2
Comparison of the number of *Crassostrea virginica* leucocytes on stained
Millipore filters from chemotaxis chambers[*]

Condition	No. of leucocytes										Mean ± S.D.
Controls	40	12	14	25	35	22	42	17	18	31	25.6 ± 10.84
Experimentals	300	280	260	182	193	221	174	271	140	120	214.1 ± 61.95

[*]Initial number of leucocytes in lower chamber = ca. 550.

INTRACELLULAR SURVIVAL AND PROPAGATION OF SYMBIONTS

As stated earlier, there have been reports that apparently certain species of bacteria, fungi, and algae can survive and/or reproduce in molluscan leucocytes (Prytherch 1940; Mackin 1951; Michelson 1961; Yonge 1936, 1953). However, in the case of fungi and algae, the conclusions arrived at by the authors have not been tested experimentally. Consequently, some doubt is warranted. Relative to the algal symbionts, this skepticism is strengthened by Mansour's (1945, 1946, 1949) reports that zooxanthellae have never been observed by him within leucocytes of tridacnid clams. In addition, more recently, Fankboner (1971), after studying zooxanthellae and in the tridacnids *Hippopus hippopus, Tridacna gigas, T. maxima*, and *T. squamosa* by electron microscopy, has concluded that the molluscan hosts' leucocytes selectively phagocytize senescent zooxanthellae from the population at the mantle edges and these are assumed to be digested intracellularly by lysosomal enzymes. Fankboner has stated that, "This process cannot be considered 'farming' as figured by earlier work, but rather the slow systematic removal and utilization of degenerate zooxanthellae from the algal population at the clam's mantle edge." This hypothesis appears to be supported by the finding of Goreau, Goreau, and Yonge (1965) involving $^{14}CO_2$ labeling, which suggests that the turnover rate of extracellular zooxanthellae in *Tridacna maxima* is slow. In other words, the host is not utilizing or removing the healthy algal cells at a rapid rate.

If Fankboner's interpretation is correct, then it may be tentatively concluded that true intracellular algal symbiosis has yet to be demonstrated and in those instances where intracellular algae have been observed in molluscan leucocytes, these are in fact degenerate or senescent cells recognized as such by host cells and engulfed for removal by intracellular degradation or by some other means. It is likely, as Fankboner has suggested, that the physicochemical nature of abnormal or moribund cells is such that they are recognized by the host's leucocytes as "nonself" and are selectively removed and destroyed. On the other hand, there is no question that extracellular algal endosymbiosis occurs in molluscs.

It is noted that the principle as applied to algal symbionts

may also hold true for the fungus *Labryinthomyxa marina*. This parasite of *C. virginica* is pathogenic, especially during warm weather
(see reviews by Ray 1954, and Ray and Chandler 1955). Although primarily, if not exclusively, an extracellular parasite in hemolymph
and tissues, it has been reported within phagocytes. In such
instances it is possible that the arrested organisms are moribund
or have become altered for some other reason so that they are chemically distinguishable and as a result become phagocytized by host
cells. Relative to this hypothesis, it is of interest to note that
in the case of parasitization of *Crassostrea virginica* by the haplo-
sporidan *Minchinia nelsoni*, occasionally we have found the protozoan
within oyster phagocytes, although the majority are extracellular
in various tissues (Fig. 3). Again, it is possible that the intracellular specimens represent chemically altered ones that are recognized as being "nonself."

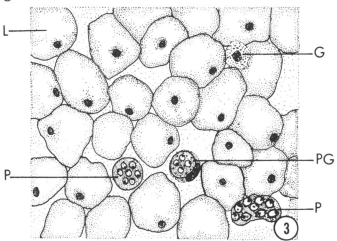

Figure 3. Drawing showing young plasmodium of *Minchinia nelsoni* within granulocyte of *Crassostrea virginica* in host's Leydig
tissue. G, granulocyte; L, Leydig cell; P, *M. nelsoni* plas-
modia; PG, phagocytized *M. nelsoni* plasmodium. (Drawn from
histological section stained with hematoxylin and eosin.)

DEGRADATION OF PHAGOCYTIZED ORGANISMS

It is known that most foreign materials phagocytized by molluscan leucocytes are usually removed either via the migration of
foreign material-laden cells across epithelial borders (see Cheng
1967 for review) and/or the nephridium (Cheng, Thakur, and Rifkin
1969) or by intracellular degradation. In recent years we have
focused our attention on the latter process and some of our findings
pertaining to two species of estuarine pelecypods, *Crassostrea
virginica* and *Mercenaria mercenaria*, are being reported at this point.
The objective of our studies is to attempt to understand the mechanisms occurring in phagocytes which prevent the intracellular establishment of apparently most potential symbionts and yet permit the
survival of others.

CRASSOSTREA VIRGINICA

Cell Types

Although a variety of cell types have been described in the hemolymph of molluscs (see Cheng and Rifkin 1970 for review), there appears to be no concensus as to exactly how many distinct categories of cells exist and what their functional roles are. Yet, such information is essential if we are to understand the processes of phagocytosis and the fate of invading organisms. In order to initiate understanding of this aspect of symbiology, we have embarked on a systematic survey of molluscan hemolymph cells. Some of our results as related to *C. virginica* have been published (Foley and Cheng 1972). In brief, as presented in Table 3, we have concluded that based on size, there are two populations of hemolymph cells in the American oyster. Among the larger cells, we have recognized granulocytes and fibrocytes. It is the granulocyte that is of particular interest from the standpoint of phagocytosis and this report includes only a discussion of this cell type.

TABLE 3
Classification of *Crassostrea virginica* hemolymph leucocytes

Cells	Types	Characteristics
Large	Granulocytes	With acidophilic granules[a] With basophilic granules[a] With refractile granules[a]
	Fibrocytes	Primary fibrocytes[b] Secondary fibrocytes[c]
Small	Hyalinocytes	Agranular Slightly granular

[a]With combination of 3 types of granules.

[b]With lobate nucleus.

[c]With spherical or ovoid nucleus.

Oyster granulocytes commonly include eosinophilic, basophilic, and refractile cytoplasmic granules when observed with the light microscope (Figs. 4-6). That a mixture of these type of granules exists in *C. virginica* granulocytes has also been recognized by Feng et al. (1971). Because of this phenomenon, Foley and Cheng (1972) have proposed that granulocytes represent one type of cell but at different ontogenetic stages. Other characteristics of these granulocytes, especially their behavior in vitro, have been reported (Foley and Cheng 1972).

Of particular interest to the topic under consideration is the fact that these granulocytes readily adhere to a solid substrate and

69

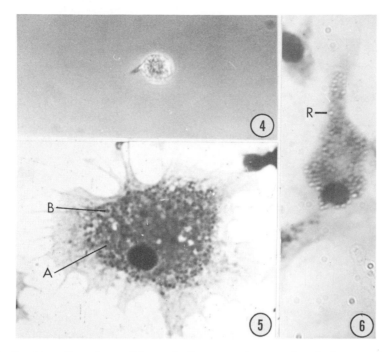

Figures 4-6. Hemolymph cells of *Crassostrea virginica*. 4. Freshly
    drawn granulocyte with single pseudopodium. 5. Granu-
    locyte fixed with 2.5 seawater glutaraldehyde, stained
    with Giemsa's solution, and photographed with a Kodak
    Wratton No. 58 green filter. 6. Granulocyte identically
    fixed and stained but photographed without the use of
    a filter showing refractile, cytoplasmic granules. A,
    acidophilic granule; B, basophilic granule; R, refrac-
    tile granule. (After Foley and Cheng 1972; with permis-
    sion of Academic Press.)

and spread (Fig. 7). There is reason to believe that this behavior
represents thigmotactic activity rather than a manifestation of
gravity. We believe that it is because of this characteristic that
our chemotaxis studies reported earlier gave erratic results. In
other words, granulocytes, which are the actively phagocytic cells
of oysters (Galtsoff 1964; Cheng and Rifkin 1970; Foley and Cheng
1972), and are the actively responding cells to invading metazoan
parasites (Rifkin and Cheng 1968; Rifkin, Cheng, and Hohl 1969),
would not be accounted for in our cell counts if they should make
contact and adhere to the sides of the lower chamber of our apparatus.
The spreading of these granular phagocytes is of interest since Bang
(1961) has reported that the mechanism of phagocytosis by oyster
leucocytes involves the entrapment of foreign materials by the cytoplasmic web between the radiating "ribs" (Fig. 8).

Electron Microscopy

We have some evidence that bacteria phagocytized by *C.virginica*
granulocytes are eventually cleared primarily as the result of the

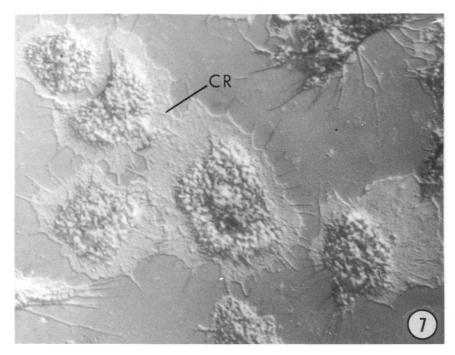

Figure 7. Spread granulocyte of *Crassostrea virginica* showing cytoplasmic granules limited to the endoplasm and the occurrence of cytoplasmic "ribs." (2.5% seawater glutaraldehyde
fixation, stained with Giemsa's solution, viewed with
Nomarski optics.) CR, cytoplasmic "rib." (After Foley and
Cheng 1972; with permission of Academic Press.)

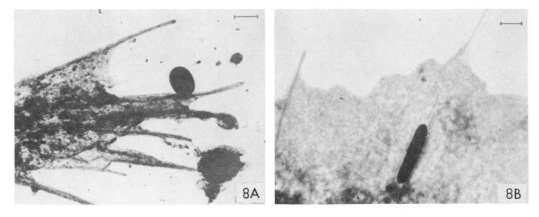

Figure 8. Electron micrographs of phagocytosis of bacteria by
Crassostrea virginica granulocyte. A, beginning of phagocytosis of uniflagellate bacterium; x 14,000. B, bacterium
taken into ectoplasm of oyster cell; x 9,000. (After Bang
1961; with permission of Biological Bulletin.)

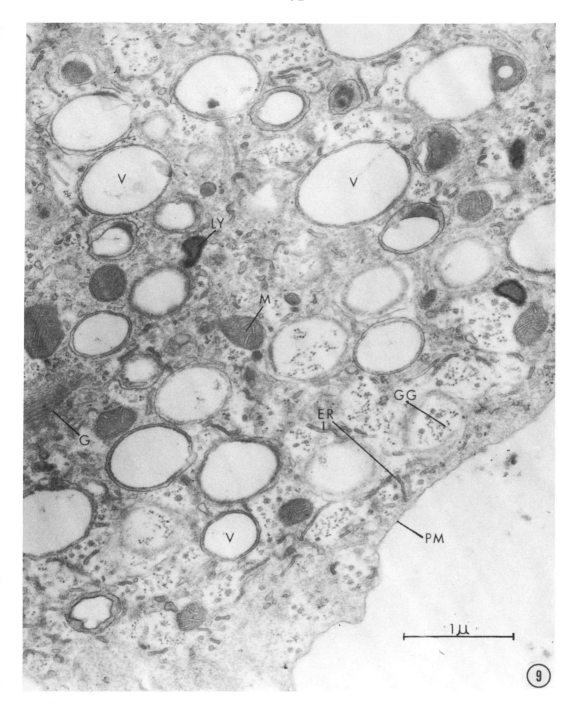

Figure 9. Electron micrograph of a portion of a granulocyte of
Crassostrea virginica. ER, endoplasmic reticulum; G,
Golgi apparatus; GG, glycogen granules; LY, lysome; M,
mitochondrion; PM, plasma membrane; V, vesicle.

migration of bacteria-laden cells across epithelial borders of the host, as has been reported by Tripp (1960). In addition, however, we also have evidence that certain bacteria are degraded intracellularly. The latter information has led us to investigate the enzyme biochemistry and fine structural architecture of these host cells. Although similar studies have been made by others (Rifkin, Cheng, and Hohl 1969; Cheng and Rifkin 1970; and Feng et al. 1971), our more recent studies have revealed some hitherto unreported details.

Granulocytes of *C. virginica* measure 13.0 ± 1.0μ by 12.0 ± 1.2μ when freshly drawn. When permitted to rest on a solid substrate, they spread in a characteristic fashion (Fig. 7) and such cells measure 36.0 ± 8.2μ by 23.5 ± 5.0μ. The nature of the granules and other constituents of these cells have been investigated with the electron microscope.

Materials and Methods

All of the oysters used were collected on bars in the proximith of Oxford, Maryland. Cells freshly drawn from the adductor muscle sinus were immediately fixed for 2 hrs in 2% glutaraldehyde in phosphate buffer at pH 7.2. After washing in phosphate buffer, the cells were post-fixed in 1% osmium tetroxide in phosphate buffer for 4 hrs, embedded in Luft's Epon, and sectioned with a glass knife

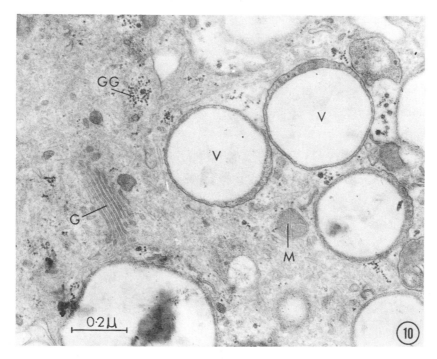

Figure 10. Electron micrograph of a portion of a granulocyte of *Crassostrea virginica* showing vesicles with cortical regions of uneven width and other cytoplasmic inclusions. G, Golgi apparatus; GG, glycogen granules; M, mitochondrion; V, vesicle.

on a Sorvall MT-2B ultramicrotome. The sections were stained with a saturated solution of uranyl acetate and lead citrate (Reynolds 1963) and viewed in a Philips 300 electron microscope operated at 60 Kv.

Results

The granulocyte of *C. virginica* is bound by a unit membrane. Mitochondria with well-formed cristae are randomly distributed throughout the cytoplasm as is smooth endoplasmic reticulum (Fig. 9). Typical stacked Golgi bodies also occur in the cytoplasm (Figs. 9, 10).

The cytoplasmic granules visible under the light microscope appear as membrane-delimited vesicles in electron micrographs. These are also randomly distributed throughout the cytoplasm. Although variations in size of cross sections and other minor differences exist, all of the vesicles share basic structural similarities. Specifically, in addition to the delimiting membrane, which is usually smooth but may be undulating (Fig. 11), there is a zone of heterogeneous medium electron density situated mediad to it (Figs. 9, 10, 12). This zone, which has been designated as the "cortex" (Feng et al. 1971), does reveal minor variations. In some vesicles it abuts the surface membrane (Figs. 9, 10, 12, 13), while in others an electron-lucid zone separates it from the membrane (Figs. 9, 11, 12). Furthermore, in some cases where the cortex is separated from the surface membrane, granules and/or globules of medium electron density occupy the lucid zone between the two layers (Fig. 10, 12, 13).

In some sections of these vesicles the cortex is of more or less uniform width, measuring 100 mμ thick; however, in other sections the cortex is thicker in one area that others (Fig. 10). This, plus the fact that all of the sections with smaller diameters have a thicker cortex, suggests that each vesicle is ovoid and the cortex at the terminals is thicker (Fig. 14). Thus, sections with nonuniformly thick cortical zones are interpreted to be tangential sections.

Of particular interest is the fact that concentrically arranged, electron-dense lamellae have been observed embedded in the cortex of some vesicles (Fig. 12). The presence of these structures causes the surface membrane to bulge towards the exterior. These bodies, as will be reported in detail elsewhere, are associated with the digestive function of these vesicles.

Mediad to the cortex is a lucid zone, designated as the "core" (Feng et al. 1971). Actually this zone appears more like a vacuolar lumen than a "core." Although this space is usually devoid of inclusions in most vesicles, electron-dense granules identifiable as glycogen granules, primarily alpha rosettes (Figs. 15, 16), occur in some and relatively large, amorphous globules of heterogeneous electron density occur in others (Fig. 17). It is possible that all of the vesicles included granules and/or the amorphous material but

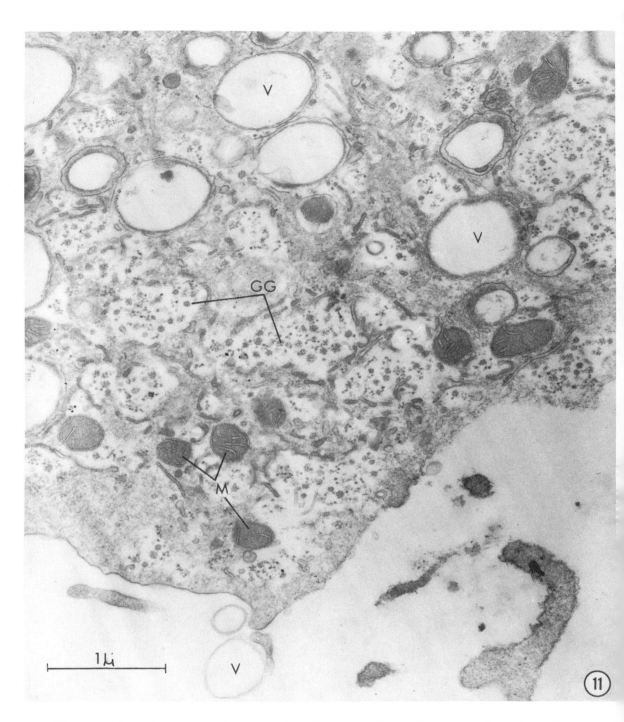

Figure 11. Electron micrograph of a portion of a granulocyte of
Crassostrea virginica showing granules arranged in
"packets" delimited by loosely arranged vesicular lamellae and other cytoplasmic inclusions. Notice the presence
of vesicles outside the granulocyte. GG, glycogen
granules; M, mitochondria; V, vesicles.

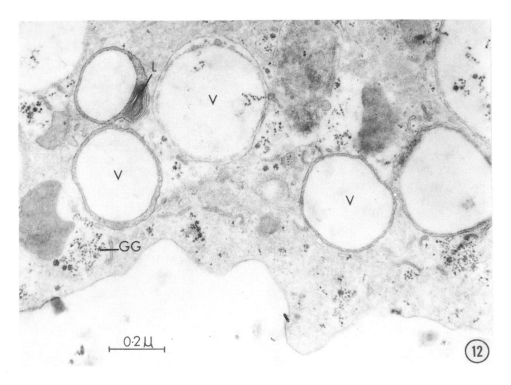

Figure 12. Electron micrograph of a portion of a granulocyte of
Crassostrea virginica showing an electron-dense concen-
tric lamellar body embedded in the cortex of a vesicle.
GG, glycogen granules; L, lamellar body; V, vesicle.

their contents were lost during fixation. It is of interest to note
that vesicles of the type described have been observed with consistency outside of granulocytes (Figs. 11, 13).

In addition to the vesicles described, lysosomes also occur in
the cytoplasm of granulocytes. These are membrane-delimited and
include electron-dense, amorphous materials (Figs. 9, 18).

Another striking feature of *C. virginica* granulocytes is the
distribution pattern of their glycogen contents. In addition to
their occurrence within vesicles, these cells also include glycogen
in the cytoplasm where both alpha rosettes and beta granules occur.
The interesting feature is that these glycogen granules appear to
be deposited in "packets" loosely enclosed by vesicular lamellae
(Figs. 9, 11). Only rarely are nonenclosed glycogen granules
observed. The "packets" of glycogen are uniformly distributed
throughout the cytoplasm. There is some evidence that this unusual
arrangement of cytoplasmic glycogen, i.e., circumscribed by loosely
arranged vesicular lamellae, is associated with the intravesicular
synthesis of this polysaccharide. Since this aspect of granulocyte
physiology is not of immediate concern to the topic at hand, the
information will be published at a later date.

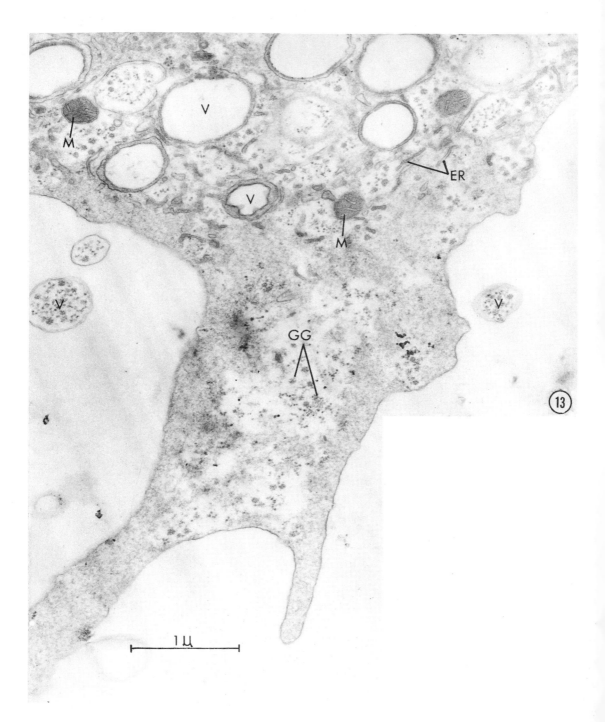

Figure 13. Electron micrograph of portion of the cytoplasm of a partially spread of *Crassostrea virginica* showing absence of vesicles (the granules of light microscopy) in the peripheral ectoplasm, the presence of vesicles in the endoplasm, and vesicles on the exterior of the granulocyte.

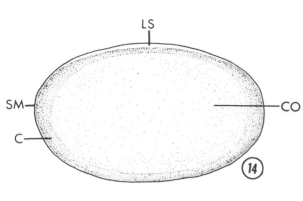

Figure 14. Schematic drawing of a single vesicle (the granule of light microscopy) from the cytoplasm of a granulocyte of *Crassostrea virginica* showing thicker cortex at the two ends and at a region where the surface membrane is lifted from the cortex, thus creating a lucid space between the surface membrane and the underlying cortex. C, cortex; CO, "core" of vesicular lumen; LS, lucid space between surface membrane and cortex; SM, surface membrane.

Discussion

Feng et al. (1971), in their electron microscope study of the granulocytes of *C. virginica*, chose to recognize three distinct types of granules (our vesicles) based on such criteria as dimensional differences of sections, thickness of the cortex, and the absence or presence of the lucid zone between the surface membrane and the cortex. Furthermore, they attempted to correlate the three types of granules as observed with the electron microscope with the acidophilic, basophilic, and refractile granules as seen with the light microscope. Although our electron micrographs have revealed the basic features reported by them, we are in disagreement relative to interpretation. We are of the opinion that all of the vesicles are of the same type. Such minor variations as thickness of the cortex and differences in cross-sectional dimensions can be explained by the plane of section readily visualized from our model (Fig. 14).

We have also observed the lucid zone between the cortex and the surface membrane associated with some vesicles; however, we do not consider this feature to be a reliable criterion for recognizing a specific type of vesicle. Rather, this lucid zone is considered to represent a separation between the surface membrane and the cortex which may be an artifact or, perhaps, a morphogenetic stage during the breakdown of vesicles. The latter is a distinct possibility since preliminary results in our laboratory indicate that the vesicular lamellae which loosely circumscribe the packets of glycogen may be formed from the reorganization of the vesicular wall. Some evidence for this is presented in the electron micrographs included herein. Specifically, well-defined glycogen granules are consistently found only within those vesicles in which the cortex is separated from the surface membrane (Figs. 9, 12, 15), or in vesicles the walls of which appear to be in the process of undergoing change (Fig. 16).

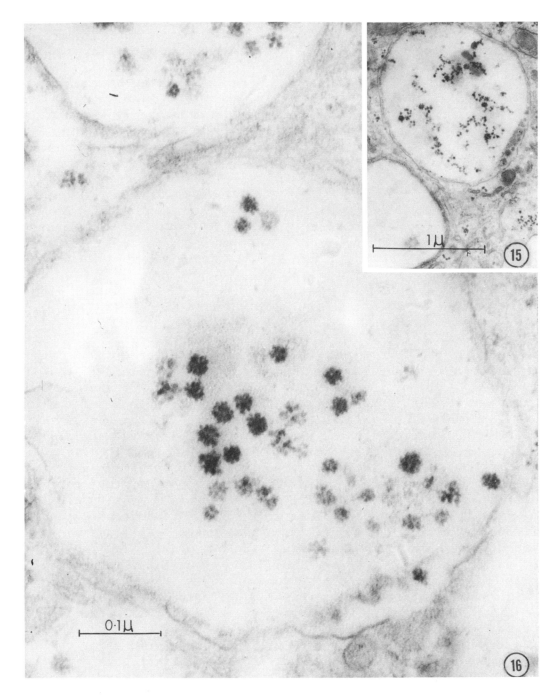

Figures 15, 16. Electron micrographs showing presence of glycogen
granules within lumina of cytoplasmic vesicles.
Notice the clearly defined alpha rosettes in Figure
16.

It is difficult for us to conceive how the diminutive thickness of the surface membrane and the cortex of each vesicle could account for the acidophilic and basophilic nature of these bodies as observed with the light microscope. In all probability the staining reactions of the granules are due to the nature of the contents of the vesicles in the living state. Studies are currently under way to test this hypothesis.

Of particular interest to the topic of host cell-foreign organism interaction is the fact that lysosomes are now known to occur in oyster granulocytes. These hydrolytic, enzyme-enclosing organelles (Novikoff 1963) are structurally distinct from the cytoplasmic vesicles (the cytoplasmic granules of light microscopy). Since certain lysosomal enzymes, especially the phosphatases, have been demonstrated in the cytoplasm of oyster leucocytes (Table 4), one is tempted to conjecture that the intracellular degradation of incompatible invading organisms or chemically altered compatible extracellular symbionts that become phagocytized is effected by lysosomal enzymes, at least in part.

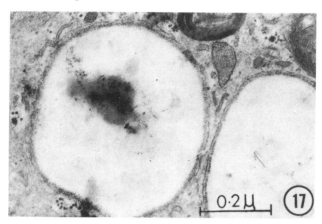

Figure 17. Electron micrograph of single vesicle in the cytoplasm of a granulocyte of *Crassostrea virginica* containing an amorphous substance of heterogeneous electron density.

Figure 18. Electron micrograph of portion of the cytoplasm of a granulocyte of *Crassostrea virginica* with a typical lysosome embedded therein. ER, endoplasmic reticulum; GG, glycogen granules; LY, lysosome; M, mitochondria; V, vesicle.

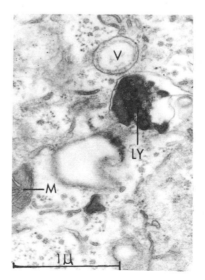

TABLE 4
Hydrolytic enzymes that have been reported from molluscan hemolymph cells

Enzymes	Mollusc	Site	Reference
Lipase	Ostrea edulis	In phagocytes	Yonge, 1926
		In amoebocytes	Takatsuki, 1934
	Crassostrea virginica	In cytoplasm of leucocytes	George, 1952
	Mercenaria mercenaria	In cytoplasm of leucocytes	Zacks and Welsh, 1953; Zacks, 1955
Protease	Ostrea edulis	In amoebocytes	Takatsuki, 1934
Amylase	Ostrea edulis	In amoebocytes	Takatsuki, 1934
Glycogenase	Ostrea edulis	In amoebocytes	Takatsuki, 1934
Enzymes active on maltose, lactose, glucosides, amygdaline,& salicine	Ostrea edulis	In amoebocytes	Takatsuki, 1934
Alkaline phosphatase	Helix aspersa	Amoebocyte nuclei	Wagge, 1951
	Crassostrea virginica	Amoebocyte cytoplasm	Cheng in Cheng and Rifkin, 1970
	C. virginica	Granulocyte cytoplasm	Feng et al., 1971
Acid phospha-tase	C. virginica	Granulocyte cytoplasm	Feng et al., 1971
Phosphomono-esterase II (acid phospha-tase)	Tridacna maxima	Vacuoles of phagocytes	Fankboner, 1971
Nonspecific esterase	C. virginica	Granulocyte cytoplasm	Feng et al., 1971

The fine structure of the leucocytes of the freshwater pulmonate *Helisoma anceps*, in which Michelson (1961) found viable and dividing acid-fast bacilli, has not yet been investigated. However, one could postulate that the reason these pathogenic bacterial symbionts are not destroyed intracellularly may be due to the lack of the necessary enzymes, or more likely, the existing enzymes are ineffective against the bacteria.

MERCENARIA MERCENARIA

Our studies on the hemolymph cells of *M. mercenaria* have thus far been limited to elucidating the cell types present and to determining which type is actively phagocytic.

Cell Types

Unlike *Crassostrea virginica*, the hemolymph cells of *M. mercenaria* are of a single population relative to size. All of the cells, when in the spread condition, i.e., extended over a solid substrate, measure from 20-30μ in greatest diameter. However, on the basis of structure, two distinct cell types have been recognized. The first, designated as granulocytes, is characterized by the presence of conspicuous granules which, as in the case of those of *C. virginica*, are limited to the endoplasm when in the spread condition (Fig. 19), but are uniformly distributed throughout the cytoplasm in freshly drawn cells.

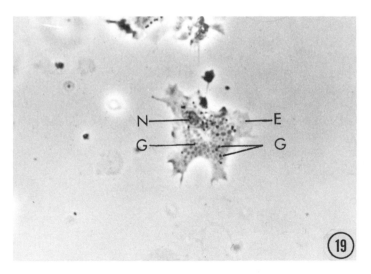

Figure 19. Photomicrograph of living granulocyte of *Mercenaria mercenaria* spread against a glass substate showing cytoplasmic granules limited to the endoplasm (40x obj.; phase-contrast optics). G, cytoplasmic granules; E, ectoplasm; N, nucleus.

When stained with Giemsa's solution at pH 6.5 after fixation with 2.5% seawater-glutaraldehyde, the cytoplasmic granules appear as a mixture of acidophilic and faintly basophilic ones. Refractile granules also occur, but these, from our experience, are only readily visible in living cells. The second type of hemolymph cells in *M. mercenaria*, designated as fibrocytes, is very similar to *C. virginica* fibrocytes (Foley and Cheng 1972).

PHAGOCYTOSIS

Materials and Methods

In order to determine whether it is the granulocyte or the fibrocyte of *M. mercenaria* that is phagocytic, the following experiment has been carried out.

A Millipore-filtered seawater suspension of 3 x 10^6/mm^3 of the large (2-3μ), gram-positive bacterium *Bacillus megaterium* freshly harvested from peptone broth and rinsed twice in sterile physiological saline was employed as the foreign agent. Initial attempts at tracing the fate of bacteria injected into clams failed to give conclusive results; therefore, the following alternative in vitro system was substituted.

A 0.02 ml sample of *M. mercenaria* hemolymph freshly drawn from the posterior adductor muscle sinus was placed on a clean 2 x 2 cm coverglass and permitted to stand for 5 min to allow the cells to adhere to the glass. The coverglass was then inverted over the well of a concave-bottom, depression slide containing 0.02 ml of the bacterial suspension. The hemolymph and the suspension coalesced so that the bacterial cells could make contact with the spread molluscan cells.

The preparations were observed continuously with phase-contrast microscopy for 2-3 hrs. In addition, coverglasses with attached hemolymph cells that had been exposed to bacterial suspensions for 25 min were subsequently fixed in 2.5% Millipore-filtered seawater-glutaraldehyde and stained with 4% Giemsa's solution at pH 6.5 for 5 min.

Results and Comments

Observations on the living in vitro preparations revealed that the bacilli commonly made contact with spread granulocytes and adhered to their surfaces. That adhesion does occur is verified by the fact that when the granulocytes moved, the bacteria attached to their surfaces also moved. On a few occasions bacilli have been observed partially embedded in the ectoplasm, although the exact mechanism of phagocytosis could not be resolved with the light microscope. Nevertheless, *B. megaterium* does become phagocytized by granulocytes since intra-cytoplasmic vacuoles containing this

this bacillus have been observed in live and stained preparations
(Figs. 20-22).

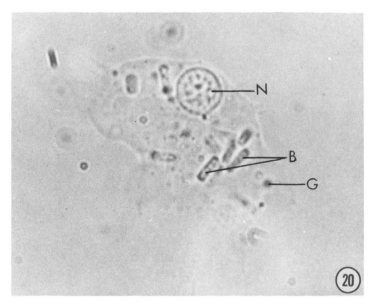

Figure 20. Photomicrograph of living granulocyte of *Mercenaria mercenaria* after exposure to *Bacillus megaterium*, showing
bacteria in cytoplasm. B, bacteria; G, cytoplasmic
granule; N, nucleus.

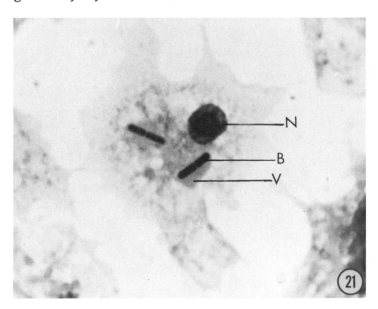

Figure 21. Photomicrograph of fixed and stained granulocyte of
Mercenaria mercenaria after exposure to *Bacillus megaterium* showing bacteria in cytoplasmic vacuoles. The
vacuoles stain acidophilically with Giemsa's solution at
pH 6.5; B, bacterium; N, nucleus of granulocyte; V,
bacteria-containing vacuole.

84

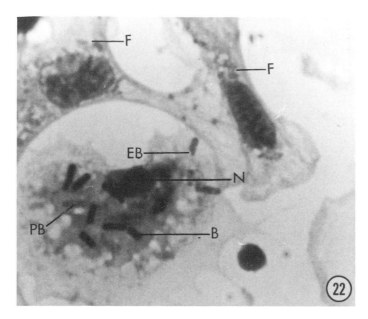

Figure 22. Photomicrograph of two fibrocytes of *Mercenaria mercenaria*
situated adjacent to a granulocyte. The cells had been
exposed to *Bacillus megaterium* and the bacteria are found
only in the granulocyte. Notice that certain bacteria are
poorly stained while others are partially eroded (2.5%
seawater-glutaraldehyde fixation; Giemsa's stain at pH
6.5). B, normal appearing phagocytized bacterium; EB,
partially eroded bacterium; F, fibrocyte; N, nucleus of
granulocyte; PB, poorly stained bacterium.

Two significant observations deserve special attention. First,
it is the granulocyte of *M. mercenaria* that is phagocytic, since
intracytoplasmic bacilli are consistently found within this type of
cell. As shown in Figure 22, even when fibrocytes and granulocytes
are situated adjacent to each other, phagocytized bacteria almost
exclusively occur in the latter. Second, although it is recognized
that morphological evidence alone is insufficient to prove the hypothesis, the consistent finding of poorly stained *B. megaterium* cells
within vacuoles of the same granulocytes that include intensely
stained bacteria suggests that the poorly stained bacteria had
become somehow altered chemically after being phagocytized (Fig. 22).
That some form of degradation had occurred is further supported by
the finding of bacilli with eroded surfaces (Fig. 23). *Bacillus
megaterium* is a free-living bacterium and in this instance, when
exposed to *M. mercenaria* granulocytes, is readily phagocytized; in
other words, it is recognized as being "nonself," and is possibly
degraded intracellularly.

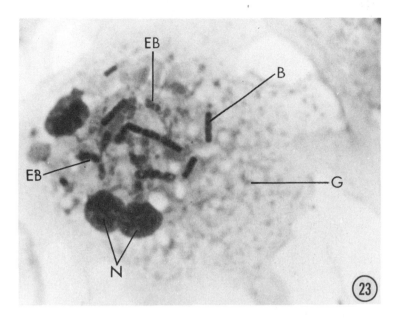

Figure 23. Photomicrograph of a multinucleate granulocyte of
Mercenaria mercenaria enclosing phagocytized *Bacillus
megaterium*. Notice several partially eroded bacteria.
The multinulceated condition of the host cell is the
result of fusion of adjacent uninucleate cells (2.5%
seawater-glutaraldehyde fixation; Giemsa's stain at pH
6.5). B, normal-appearing bacterium; EB, partially
eroded bacteria; G, cytoplasmic granule; N, nuclei of
host cell.

INTRACELLULAR DEGRADATION AND HOST UTILIZATION

Our finding of lysosomes in the phagocytic granulocytes of *C.
virginica* and Fankboner's (1971) report of these organelles in
"amoebocytes" of *Hippopus hippopus* raise the possibility that the
intracellular degradation of phagocytized materials may be associated
with the action of lysosomal enzymes. However, direct evidence for
the occurrence of these hydrolytic enzymes is scanty. Yet, since
these cells are responsible for the intracellular degradation of
organisms, it would appear to be of considerable interest to know
what enzymes occur, how specific these are to substrates, and what
their optimal ranges are. Such information obviously is essential
if we are to understand in depth the role of host cells in endosymbiosis. In Table 4 are listed those hydrolytic enzymes that have
been reported from molluscan leucocytes.
The question may be raised at this point as to whether the
intracellular digestion of phagocytized organisms is merely a mechanism for the elimination of foreign bodies or does it also serve the

the host to some other advantage. Smith, Muscatine, and Lewis (1969) have reviewed what is known about the utilization of symbiotic algae as a holozoic food source by their invertebrate hosts. Although intracellular symbionts of some cnidarians are believed to benefit their hosts metabolically, especially from the standpoint of nutrition, there is yet no definitive evidence that cells degraded within molluscan phagocytes are used as a nutrient source. On the other hand, it is now quite apparent that extracellular symbionts of marine molluscs, e.g., the algae in the siphonal hemal sinuses of tridacnids, are true mutualists.

GENERAL DISCUSSION AND CONCLUSIONS

From the available information, it appears that the granulocytes of at least two species of marine molluscs, *Crassostrea virginica* and *Mercenaria mercenaria*, are extremely efficient in recognizing "self" from "nonself" and foreign materials small enough to be phagocytized are readily arrested by this mechanism. The same appears to be true in the gastropods *Lehmania poirieri* (Arcadi 1968), *Littorina scabra* (Cheng, Thakur, and Rifkin 1969), and *Aplysia californica* (Pauley and Krassner 1972), although the specific type of cell involved in phagocytosis has yet to be identified in these gastropods.

Phagocytized "nonself" organisms may survive, but this may be the result of the ineffectiveness of the intracellular enzymes of phagocytes against such organisms. However, most known organisms that become phagocytized are eliminated or intracellularly degraded. This does not mean that microbial and other small symbionts do not occur in molluscs; however, in the case of *Minchinia nelsoni*, *Labyrinthomyxa marina*, the zooxanthellae and zooxanthellae occurring in various molluscs, and other known symbionts, these are intercellular and unless they become altered in some manner, are recognized as "self" and are sustained within the molluscs' bodies.

The only known successful intracellular symbionts of marine molluscs, as far as we have been able to determine, are certain telosporean protozoa such as certain species of *Nematopsis*, *Porospora*, *Pseudoklossia*, and *Hyaloklossia*; the haplosporean *Haplosporidium tumefacientis*; and the microsporidans *Chytridiopsis mytilovum* and *C. ovicola*. These occur within phagocytes or within the cells of the gills, digestive tract, or nephridium (Cheng 1967). Relative to those in the ctenidia and digestive tract, it could be argued that direct penetration from the exterior (i.e., ctenidial surfaces and gastric or intestinal lumen) does not allow for any significant contact between the symbiont and the molluscan host's phagocytes, and consequently, these are the most vulnerable host cells for invasion. Again, since it has been shown that the nephridium is one of the primary sites for the deposition of foreign materials introduced into the bodies of certain molluscs (Cheng, Thakur, and Rifkin 1969), it could be reasoned that protozoan parasites occurring at this site have been carried there by phagocytes.

In the case of *Nematopsis ostrearum*, it is the spore that occurs within phagocytes and it is probably protected by the spore wall from degradation. In fact, Feng (1958) has expressed the opinion that the spores of *N. ostrearum* in oyster phagocytes are in the process of being expelled and hence are only temporarily within these cells.

In support of our general thesis, it could be mentioned that although the Microsporida are widely distributed as intracellular parasites of invertebrates (Kudo 1924), they are rarely found in molluscs. Even those species that have been reported from molluscs occur as hyperparasites in trematode sporocysts and rediae, e.g., *Nosema echinostomi* in echinostome rediae within *Lymnaea limosa* (Brumpt 1922), *Nosema dollfusi* in *Bucephalus cuculus* sporocysts within *Crassostrea virginica* (Sprague 1964), *Perezia helminthorum* in trematode larvae in Malayan snails (Canning and Basch 1968), the unidentified microsporidan in Malayan echinostome rediae (Lie, Basch, and Umathevy 1966), the unidentified species reported by Schäller (1959) in larval trematodes in *Tropidiscus planorbis*, and *Nosema strigeoideae* in the intramolluscan stages of *Diplostomum flexicaudum* in *Stagnicola emarginata angulata* (Hussey 1971). As hyperparasites, they are protected from the phagocytic action of the molluscan hosts' hemolymph cells. The same holds true for the microsporidans *Chytridiopsis mytilovum* and *C. ovicola* reported from within the ova of *Mytilus edulis* and *Ostrea edulis*, respectively (Field 1924; Sprague 1963; Léger and Hollande 1917).

It is the nature of biological principles that exceptions invariably occur. Consequently, what appears to be the basis for our thesis may not be supported by subsequent findings as symbiologists in increasing numbers direct their attention to the mollusc as a host and examine the host-symbiont interaction as they have during the past decade.

ACKNOWLEDGMENTS

The original data included in this article have resulted from research supported by Grant FD-00416-01 from the U. S. Public Health Service. This paper is Contribution No. 103 from the Center for Marine and Environmental Studies, Lehigh University.

LITERATURE CITED

Arcadi, J. A. 1968. Tissue response to the injection of charcoal into the pulmonate gastropod *Lehmania poirieri*. J. Invertebr. Pathol. 11:59-62.

Bang, F. B. 1961. Reaction to injury in the oyster (*Crassostrea virginica*). Biol. Bull. 121:57-68.

Brumpt, E. 1922. Précis de Parasitologie. 3rd ed. Paris: Masson.

Buchner, P. 1965. Endosymbiosis of animals with plant microorganisms. New York: Interscience.

Canning, E. W., and Basch, P. F. 1968. *Perezia helminthorum* sp. nov., a microsporidian hyperparasite of trematode larvae from Malaysian snails. Parasitology 58:341-47.

Cheng, T. C. 1966. Perivascular leucocytosis and other types of cellular reactions in the oyster *Crassostrea virginica* experimentally infected with the nematode *Angiostrongylus cantonensis*. J. Invertebr. Pathol. 8:52-58.

_____. 1967. Marine molluscs as hosts for symbioses. In:Advances in marine biology, ed. F. S. Russell, vol. 5. London and New York: Academic Press.

_____, and Cooperman, J. C. 1964. Studies on host-parasite relationships between larval trematodes and their hosts. V. The invasion of the reproductive system of *Helisoma trivolvis* by the sporocysts and cercariae of *Glypthelmins pennsylvaniensis*. Trans. Am. Microsc. Soc. 83:12-23.

_____, and Rifkin, E. 1970. Cellular reactions in marine molluscs in response to helminth parasitism. In: A symposium on diseases of fishes and shellfishes, ed. S. F. Snieszko, pp. 443-96. Washington, D. C.: Amer. Fisher. Soc., Spec. Publ. No. 5.

_____, and Sanders, B. G. 1962. Internal defense mechanisms in molluscs and an electrophoretic analysis of a naturally occurring serum hemaglutinin in *Vivparus malleatus* Reeve. Proc. Pa. Acad. Sci. 36:72-83.

_____, Shuster, C. N., Jr., and Anderson, A. H. 1966a. A comparative study of the susceptibility and response of eight species of marine pelecypods to the trematode *Himasthla quissetensis*. Trans. Am. Microsc. Soc: 85:284-95.

_____, _____, and _____. 1966b. Effects of plasma and tissue extracts of marine pelecypods on the cercaria of *Himasthla quissetensis*. Exptl. Parasitol. 19:9-14.

_____, Thakur, A. S., and Rifkin, E. 1969. Phagocytosis as an internal defense mechanism in the Mollusca: with an experimental study of the role of leucocytes in the removal of ink particles in *Littorina scabra* Linn. In: Proceedings of the symposium on Mollusca, Part II, pp. 546-63. Marine Biol. Assoc. India. Bangalore, India: The Bangalore Press.

Dubois, R. 1901. Sur le mécanisme de la formation des perles fines dans le *Mytilus edulis*. C. R. Habd. Séanc. Acad. Sci., Paris 133:603-605.

_____. 1907. Action de la chaleur sur le distome immaturé de *Gymnophallus margaritarum*. C. R. Séanc. Soc. Biol. 63:502-504.

Fankboner, P. V. 1971. Intracellular digestion of symbiontic zooxanthellae by host amoebocytes in giant clams (Bivalvia: Tridacnidae), with a note on the nutritional role of the hypertrophied siphonal epidermis. Biol. Bull. 141:222-34.

Feng, S. Y. 1958. Observations on distribution and elimination of spores of *Nematopsis ostrearum* in oysters. Proc. Natl. Shellfish Assoc. 48:162-73.

_____. 1959. Defense mechanism of the oyster. Bull. N. J. Acad. Sci. 4:17.

_____. 1965. Pinocytosis of proteins by oyster leucocytes. Biol. Bull. 128:95-105.

_____. 1966. Experimental bacterial infection in the oyster *Crassostrea virginica*. J. Invertebr. Pathol. 8:505-11.

_____. 1967. Responses of molluscs to foreign bodies, with special reference to the oyster. Fed. Proc. 26:1685-92.

_____, Feng, J. S., Burke, C. N., and Khairallah, L. H. 1971. Light and electron microscopy of the leucocytes of *Crassostrea virginica* (Mollusca:Pelecypoda). Z. Zellforsch. 120:222-45.

Field, I. A. 1924. Biology and economic value of the sea mussel *Mytilus edulis*. Bull. U. S. Bur. Fish. 38:128-259.

Foley, D. A., and Cheng, T. C. 1972. Interaction of molluscs and foreign substances: The morphology and behavior of hemolymph cells of the American oyster, *Crassostrea virginica*, in vitro. J. Invertebr. Pathol. 19:383-94.

Galtsoff, P. S. 1964. The American oyster, *Crassostrea virginica* (Gmelin). Fish. Bull. Fish Wildlife Serv. U. S. 64:1-480.

George, W. C. 1952. The digestion and absorption of fat in lamellibranchs. Biol. Bull. 102:118-27.

Giard, A. 1907. Sur les trématodes margaritigènes du Pas-de-Calais (*Gymnophallus somateriae* Levinson et *G. bursicola*(Odhner). C. R. Séanc. Soc. Biol. 63:416-20.

Goetsch, W., and Scheuring, L. 1926. Parasitismus und Symbiose der Algengattung *Chlorella*. Zeit. Morph. Okol. Tiere 7:220-53.

Goreau, T. F., Goreau, N. I., and Yonge, C. M. 1965. Evidence for a soluble algal factor produced by the zooxanthellae of *Tridacna elongata* (Bivalvia, Tridacnidae). (Unpublished manuscript available from the Department of Zoology, University of Edinburgh, West Mains Road, Edinburgh, Scotland.)

Hussey, K. L. 1971. A microsporidan hyperparasite of strigeoid trematodes, *Nosema strigeoideae* sp. n. J. Protozool. 18:676-79.

Jameson, H. L. 1902. On the origin of pearls. Proc. Zool. Soc. London 1:140-66.

Kudo, R. 1924. A biological and taxonomic study of the Microsporidia. Ill. Biol. Monogr. 9:1-268.

Léger, L., and Hollande, A. C. 1917. Sur un nouveau protiste a facies de *Chytridiopsis*, parasite des ovules de l'huître. C. R. Séanc. Soc. Biol. 80:61-64.

Lie, K. J., Basch, P. F., and Umathevy, T. 1966. Studies on Echinostomatidae (Trematoda) in Malaya. XII. Antagonism between two species of echinostome trematodes in the same lymnaeid snail. J. Parasitol. 52:454-57.

Mackin, J. G. 1951. Histopathology of infection of *Crassostrea virginica* by *Dermocystidium marinum*. Bull. Mar. Sci. Gulf. Caribb. 1:72-87.

Mansour, K. 1945. The zooxanthellae, morphological peculiarities and food and feeding habits of the Tridacnidae with reference to other lamellibranchs. Proc. Egypt. Acad Sci. 1:1-11.

———. 1946. Source and fate of the zooxanthellae of the visceral mass of *Tridacna elongata*. Nature 158:130.

———. 1949. The morphological and biological peculiarities of *Tridacna elongata* and *T. squamosa*. C. R. 13th Congr. Intl. Zoology, Paris, 1948:441-44.

Michelson, E. H. 1961. An acid-fast pathogen of fresh-water snails. Am. J. Trop. Med. Hyg. 10:423-33.

Novikoff, A. B. 1963. Lysosomes in the physiology and pathology of cells: contributions of staining methods. In: Lysosomes, eds. A. V. S. de Reuck, M. P. Cameron, pp. 36-73. Ciba Foundation Symposium. Boston, Massachusetts: Little, Brown and Company.

Pauley, G. B., and Krassner, S. M. 1972. Cellular defense reactions to particulate materials in the California sea hare, *Aplysia californica*. J. Invertebr. Pathol. 19:8-17.

Perrier, E. 1903. Remarques de M. Edm. Perrier àpropos de la communication de M. Raphaël Dubois, de 19 Octobre dernier, "sur les huîtres perliéres vraies." C. R. Hebd. Séanc. Acad. Sci., Paris 137:682.

Prytherch, H. F. 1940. The life cycle and morphology of *Nematopsis ostrearum* sp. nov., a gregarine parasite of the mud crab and oyster. J. Morph. 66:39-64.

Ray, S. M. 1954. Biological studies of *Dermocystidium marinum*, a fungus parasite of oysters. Rice. Inst. Pamph. Special Issue, Nov., 1954. Houston, Texas: The Rice Institute.

———, and Chandler, A. C. 1955. *Dermocystidium marinum*, a parasite of oysters. Exptl. Parasitol. 4:172-200.

Read, C. P. 1971. The microcosm of intestinal helminths. In: Ecology and physiology of parasites, ed. A. M. Fallis, pp. 188-200. Toronto, Canada: Univ. of Toronto Press.

Reynolds, E. S. 1963. The use of lead citrate at high pH as an electron opaque stain in electron microscopy. J. Cell Biol. 17:208-12.

Rifkin, E., and Cheng, T. C. 1968. The origin, structure, and histochemical characterization of encapsulating cysts in the oyster *Crassostrea virginica* parasitized by the cestode *Tylocephalum* sp. J. Invertebr. Pathol. 14:54-64.

———, ———, and Hohl, H. R. 1969. An electron microscope study of the constituents of encapsulating cysts in *Crassostrea virginica* formed in response to *Tylocephalum* metacestodes. J. Invertebr. Pathol. 14:211-26.

Schäller, G. 1959. Microsporidienbefall und Degenerationserscheinungen der Trematodenlarven im Swischenwirt (*Tropidiscus planorbis*). Z. Wis. Zool. 162:144-90.

Smith, D., Muscatine, L., and Lewis, D. 1969. Carbohydrate movement from autotrophs to heterotrophs in parasitic and mutualistic symbiosis. Biol. Rev. 44:17-90.

Sprague, V. 1963. Revision of genus *Haplosporidium* and restoration of genus *Minchinia* (Haplosporidia, Haplosporidiidae). J. Protozool. 10:263-66.

_____. 1964. *Nosema dollfusi* N. sp. (Microsporidia, Nosematidae), a hyperparasite of *Bucephalus cuculus* n. *Crassostrea virginica*. J. Protozool. 11:381-85.

Stauber, L. A. 1950. The fate of India ink injected intracardially into the oyster, *Ostrea virginica* Gmelin. Biol. Bull. 98: 227-41.

_____. 1961. Immunity in invertebrates; with special reference to the oyster. Proc. Natl. Shellfish. Assoc. 50:7-20.

Takatsuki, S. 1934. On the nature and functions of the amoebocytes of *Ostrea edulis*. Quart. J. Microsc. Sci. 76:379-431.

Tripp, M. R. 1958a. Disposal by the oyster of intracardially injected red blood cells of vertebrates. Proc. Natl. Shellfish. Assoc. 48:143-47.

_____. 1958b. Studies on the defense mechanism of the oyster. J. Parasitol. 44 (Sect. 2):35-36.

_____. 1960. Mechanisms of removal of injected microorganisms from the American oyster, *Crassostrea virginica* (Gmelin). Biol. Bull. 119:210-23.

_____. 1961. The fate of foreign materials experimentally introduced into the snail *Australorbis glabratus*. J. Parasitol. 47: 745-51.

Tsujii, T. 1960. Studies on the mechanisms of shell- and pearl-formation in Mollusca. J. Faculty Fish. Prefectural Univ. Mie 5:1-70.

Wagge, L. E. 1951. The activity of amoebocytes and of alkaline phosphatases during the regeneration of the shell in the snail *Helix aspersa*. Quart. J. Microsc. Sci. 92:307-21.

Yonge, C. M. 1926. Structure and physiology of the organs of feeding and digestion in *Ostrea edulis*. J. Mar. Biol. Ass. U. K. 14:295-388.

_____. 1936. Mode of life, feeding, digestion and symbiosis with zooxanthellae in the Tridacnidae. Sci. Rep. Gr. Barrier Reef Exped. 1:283-321.

_____. 1953. Mantle chambers and water circulation in the Tridacnidae (Mollusca). Proc. Zool. Soc. London 123:551-61.

Zacks, S. I. 1955. The cytochemistry of the amoebocytes and intestinal epithelium of *Venus mercenaria* (Lamellibranchiata), with remarks on a pigment resembling ceroid. Quart. J. Microsc. Sci. 96:57-71.

_____, and Welsh, J. H. 1953. Cholinesterase and lipase in the amoebocytes, intestinal epithelium and heart muscle of the quahog, *Venus mercenaria*. Biol. Bull. 105:200-11.

Double infections of larval trematodes: competitive interactions

Patricia J. DeCoursey and Winona B. Vernberg
Belle W. Baruch Coastal Research Institute
University of South Carolina
Columbia, South Carolina

The mud snail *Nassarius obsoletus* (Say) is a prosobranch gastropod occurring in vast numbers on the extensive mudflats along the eastern coast of the United States. Along the Carolina coast, *Nassarius* is the intermediate host for at least nine species of larval trematodes. All are digenetic trematodes with remarkably complex life histories, alternating a free-living state with a highly specific obligate parasitic one. A common sequence in the life cycle of trematode species involves egg, free-living miracidium, intramolluscan sporocyst or redial stage, free-living cercarial stage, parasitic metacercarial stage in a poikilothermic second intermediate host, and finally a parasitic adult worm in the definitive vertebrate host (Fig. 1; and Cheng 1964).

The sporocyst or redial generations of the trematodes used in this study all closely resemble each other in general structure. Moreover, all are restricted to the same general region of *Nassarius*. The snail host is relatively simple in structure, consisting anteriorly of a large retractable foot with the associated mantle, antennae, proboscus, gills, kidney, and orifices of the digestive tube. Posteriorly in the shell-covered apical coil are found the U-shaped digestive tube, the gonads, the hepatopancreas or digestive gland, and associated connective tissue storage depots. It is in this posterior region of *Nassarius* that the sporocyst-redial generation develops and produces enormous numbers of cercariae. Comparable infections are known in a variety of marine molluscs (Cheng 1962, James 1965, Rees 1936, Robson and Williams 1971).

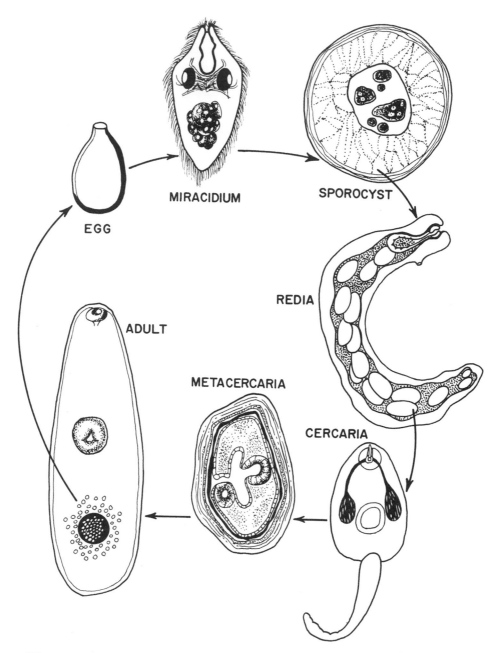

MIRACIDIUM

SPOROCYST

EGG

REDIA

ADULT

METACERCARIA

CERCARIA

Figure 1. Schematic life cycle of a digenetic trematode.

The incidence of infection with trematodes is relatively low in South Carolina populations of *N. obsoletus*. Most are single infections, but double infections do occur (Vernberg, Vernberg, and Beckerdite 1969). These double infections, living together in the relatively homogeneous microenvironment of the gonad-hepatopancreas, seem to contradict a basic concept of ecological thought concerning interspecific competition. The Gausian principle or principle of competitive exclusion states that in the competition between two species occupying the same niche, the better adapted will eventually

displace the other species. The concept has been considered extensively in terms of free-living populations (Ayala 1972; den Boer and
Gradwell 1971; Harger 1972; Lack 1971; Levin 1972). While the principle has also been applied to some parasitic helminth populations
(Holmes 1961; Schad 1963, 1966), only rarely have digenetic trematodes been considered in these terms (Lauckner 1971).

Do these species of larval trematodes truly coexist? Several
techniques have been used in our laboratory in a multifaceted
approach to this question. Field data on the incidence of infection,
collected from Beaufort, North Carolina, and Georgetown, South Carolina,
populations during the past twelve years, suggest that some combinations of species certainly do not coexist (Fig. 2). Species inci-
dences are low and slight differences between areas occur; cracking
data predictably give higher and more accurate results than shedding
data. It is also evident that three species, *Lepocreadium setiferoides*, *Zoögonus lasius*, and *Himasthla quissetensis*, are the most
abundant species in the single infections; the remaining six species
are uncommon to rare. While two of the most abundant singles, *Zoögonus* and *Lepocreadium*, are also found most commonly in the double
infections, the very frequent *Himasthla* is rare or absent in double
infections. Furthermore, some of the species relatively rare in
single infections appear with surprisingly high incidence in doubles
(Fig. 2-B). Pairing of parasites in the double infections does not,
therefore, seem to be a random process based on mere availability of
parasites.

Even more significant than the percentage of occurrence in
doubles are the actual pairing combinations. Figure 3 summarizes in
schematic form the incidence of pairing for the twenty-six double
infections available from our laboratory. Two species, *Zoögonus* and
Austrobilharzia, combine with the greatest variety of species. Of
the two, *Zoögonus* is certainly the dominant combiner, and the pair
Zoögonus-Lepocreadium alone comprises 57% of all double infections.
Based on availability, the pair *Lepocreadium-Himasthla* should be
equally abundant, but it has never been detected in our collections.

Other relationships become clear from Figure 3. Pairing does
not seem to be related to the definitive host of the trematode. While
the definitive hosts of the most common double, *Zoögonus-Lepocreadium*,
are both fish, and for *Austrobilharzia-Himasthla* are both birds,
bird-turtle and bird-fish combinations do occur.

Reports of other workers on the frequency of multiple larval
trematode infections confirm the hypothesis of nonrandom combinations.
Martin (1955) showed typical abundance patterns throughout the year
for single infections of seventeen species of larval trematodes infes-
ting the marine snail *Cerithidea californica*. The double and triple
infections did not conform to the distribution expected from random
combination, suggesting possible antagonistic interaction and elimination of some pairs. In analyses of *Nassarius* larval trematode infections, the predominant *Zoögonus lasius* was found in all but two of
fourteen double infections; significantly, the common echinostome,
Himasthla, was very rare in double infections (Vernberg, Vernberg,
and Beckerdite 1969). Similar results are reported for freshwater

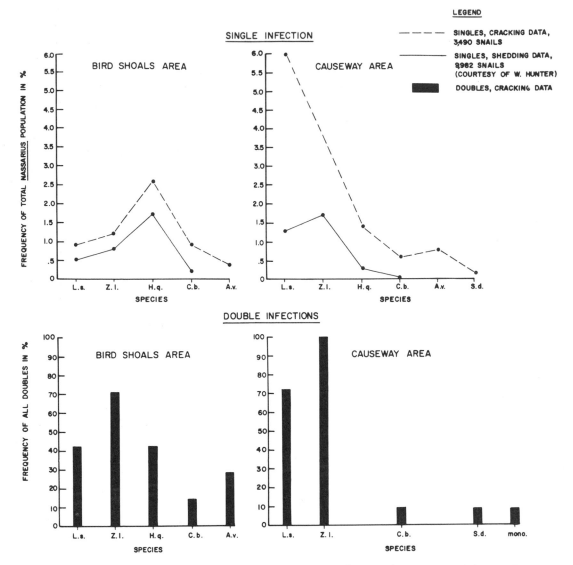

Figure 2. Incidence of single infections (above) and double infections (below) of a larval trematode in *Nassarius obsoletus*
hosts from two sites, Bird Shoals (left) and Causeway Area
(right) near Beaufort, North Carolina. For species abbre-
vations see Table 1.

snails (Bourns 1963; Heyneman and Umathevy 1968).

Often in the double infections one species suppresses cercarial
release of the other one. In our laboratory, the usual isolation
techniques did not always detect double infections; subsequent cracking procedures or histological examination of supposedly single infections occasionally revealed mature cercariae of two species. In
analyzing the available shedding data (Fig. 3), species were ranked

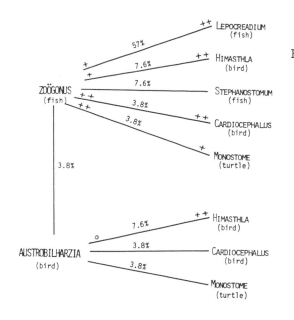

Figure 3. Summary of all pairing combinations in double infections of larval trematodes from *Nassarius obsoletus*, Beaufort, North Carolina and Georgetown, South Carolina, 1960-72. Taxonomic group in parenthesis states the definitive host for adult trematode worm; % = incidence of infection; + and O for suppression of shedding.

from O (completely suppressed) to ++ (uninhibited). Even in the most common double pair, *Zoögonus-Lepocreadium*, *Lepocreadium* appeared to suppress shedding of *Zoögonus*.

An experiment was carried out in our laboratory to test in more quantitative terms the shedding in *Zoögonus-Lepocreadium* pairs. Highly specific daily patterns of cercarial shedding have been reported in the literature (Craig 1972; Valle, Pellegrino, and Alvarenga 1971). For this reason both the rhythmic pattern of release throughout the day as well as the daily release totals were recorded for several single infections of *Zoögonus* and *Lepocreadium*, and for several *Zoögonus-Lepocreadium* double infections. Freshly caught infected snails were kept singly in jars in a constant environment chamber in 30 o/oo seawater at 25C, on a light schedule of 12L:12D. At 3-hr intervals the snails were transferred to fresh seawater, without change in the light conditions, and counts of the cercariae in the complete sample were made with the aid of a dissecting scope. Both *Zoögonus* and *Lepocreadium* are predominatly day shedding species, with a tendency for *Zoögonus* to shed late in the light period (Fig. 4). While little change in temporal distribution of shedding occurred in the double infections, the pattern and daily total cercarial counts demonstrate the great suppression of *Zoögonus* shedding by *Lepocreadium* (Fig. 4).

A third clue to regulation mechanism comes from histological examination of the tissue of snails. Again, this source of information suggests that not all pairs are equally compatible. For these studies about eighty infected snails were dissected under 32X magnification to determine the indentity, extent, and distribution of the infections. Thirty snails, including representative examples of singles, doubles, and nonparasitized snails, were fixed in Bouin's solution, serially sectioned at 8 μ, and finally stained in hematoxylin and eosin. To aid in orientation, some snails were pinned in a

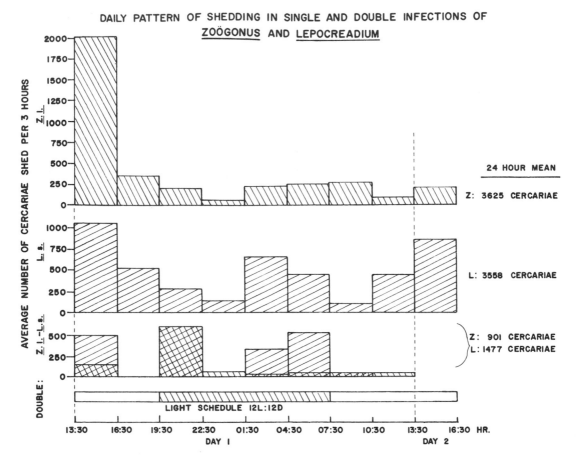

Figure 4. Daily shedding rhythms of single and double infections of
larval trematodes in *Nassarius obsoletus*: pooled data for
three specimens of *Zoögonus* (above), five *Lepocreadium*
(middle), and two *Zoögonus-Lepocreadium* (below), in 12L:12D
light cycle, 25C temperature.

straightened position before fixation; subsequent sectioning indicated
the true linear distribution in the snail tissues without the compli-
cations of a coiled apex.

    Internally the posterior half of the snail is largely comprised
of digestive and reproductive systems (Fig. 5). The gut, extremely
simple in construction, consists of a tubular loop extending from
mouth to anus. The wall is a simple epithelium varying from cuboidal
to ciliated columnar. No extensive muscular or supporting connective
tissue is found underlying the simple mucosa. Surrounding the gut is
a layer of fat and glycogen-storing cells, the Leydig cells. A large
portion of the coil is filled with tubules of the digestive gland.
Adjacent to it, extending out to the extreme tip are the gonadal
tissues. In the breeding season, approximately from December to
April (Sastry 1971), the gonads contain developing and mature gametes.
During the remainder of the year the gonads regress markedly and seem

99

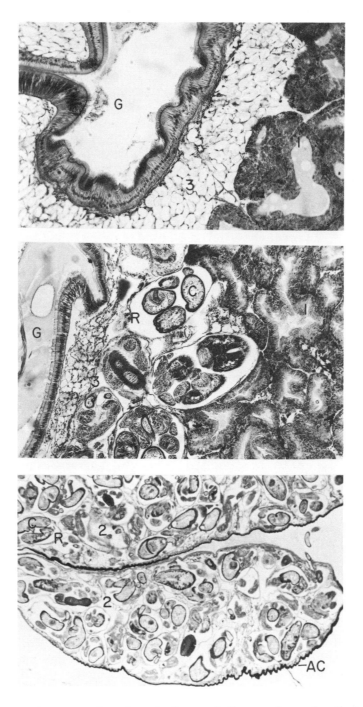

Figure 5. Photomicrograph of uninfected *Nassarius obsoletus* (above),
and snail host infected with *Himasthla quissetensis* rediae
and cercariae (middle and lower). Numbers indicate infection sites: (1) hepatopancreas, (2) gonadal area, (3) Leydig tissue surrounding gut. G=gut, AC=apical coil, R=redia,
and C=cercaria. Magnification: 120X (above), 135X (middle),
65X (below).

to be largely replaced with Leydig tissue. The relationship of parasitism to the reproductive cycle is not yet clear. Parasitic castration is well documented in marine snails (Cheng and Cooperman 1964; Cheng 1967; Rees 1936; Robson and Williams 1971). In *Nassarius*, however, it is not known whether parasites gain access and destroy an active gonad, or whether infection occurs during the sexually immature or inactive phases of the snail's development, with subsequent inhibition of normal gametogenesis by the parasites.

Details of structure are known for a variety of molluscan species parasitized by larval trematodes (Cheng and Rifkin 1970; Cheng and Snyder 1962; Cheng and Yee 1968; James 1965; James and Bowers 1967; Porter 1970). In single infections of *N. obsoletus* which we have examined, the parasites are concentrated in the hepatopancreas and gonadal areas. Anterior to this region the parasites utilize the Leydig cells, following the gut down to the anterior end of the snail. Although the exact location is influenced by the extent and age of the infection, nonetheless some preferences are discernible (Fig. 5). In spite of the simplicity of the tissues involved, four distinct niches are available. Area 1. The digestive gland. Parasites may either be distributed throughout, or restricted to some part such as the outer rim. Area 2. The gonadal area. Since there is no sharp boundary between digestive area and gonadal area, this distinction is somewhat arbitrary. Especially in very heavy infections, destruction of digestive gland tissue together with parasitic castration compounds the difficulties. However, certain areas such as the tip of the coil, the Leydig cells lateral to the gonad, or the outer rim of the coil are clearly recognizable as gonadal (Fig. 5). Area 3. The sheath of Leydig cells surrounding the gut adjacent to the hepatopancreas. Area 4. The exit route along the gut, extending anteriorly to the gut orifices.

The distribution of *Zoögonus* is relatively variable. In some infections it is primarily in the digestive gland, penetrating between the tubules, and is totally lacking in the apex of the coil. In others, particularly in heavy infections, it extends into the gonadal area. *Lepocreadium*, clearly a parasite of the gonadal area, is always found in large numbers in the apex and lateral to the hepatopancreas; large masses follow the gut anteriorly. In heavy infections it also penetrates the hepatopancreas. The monostome is exclusively a gonadal parasite. The cercariae exit at a very early stage of development from the rediae, and migrate anteriorly along the gut in the Leydig tissue; there the mature cercariae can be readily distinguished by their three eyespots and prominent cystogenous glands. The gymnocephalus type and *Stephanostomum* are almost identical in distribution, concentrated between the tubules of the digestive gland, often avoiding the rim. They may also partially invade the gonadal area. *Himasthla* appropriates the gonadal area, avoiding even in heavy infections all but the outer edge of the digestive gland. Unlike any of the other species, it is also found concentrated in the layer of Leydig cells surrounding the gut. *Austrobilharzia* penetrates between the hepatopancreas tubules and completely fills the gonadal areas. It differs from such species as the gymnocephalus type and *Stephano-stomum* in the large size of its sporocysts, which appear to exert much mechanical pressure, often obliterating the digestive gland tubules. The large extent of the gonadal area suggests that this is the prime area of concentration for *Austrobilharzia*. In contrast, *Cardiocephalus* is localized within the hepatopancreas; very little gonadal area is present, and few sporocysts are found there.

Thus, marked overlap in territory and habitat preferences occurs, but only the gymnocephalus type and *Stephanostomum* are truly identical in all choices (Table 1). It should be remembered that

TABLE 1
Relative distribution of species of larval trematodes in their first intermediate host, *Nassarius obsoletus*

Species	Hepatopancreas	Gonadal	Storage depots around gut	Along gut, anterior to Hepatopancreas
Zoögonus lasius	H	I	O	I
Lepocreadium setiferoides	I	H	O	H
Monostome cercaria	O	H	O	H
Gymnocephalus-type cercaria	H	I	O	I
Stephanostomum dentatum	H	I	O	I
Himasthla quissetensis	O	H	H	I
Austrobilharzia variglandis	I	H	O	I
Cardiocephalus brandesii	H	O	O	I

H=heaviest concentration; I=intermediate concentration; O=no parasites.

these distinctions are based solely on morphological characteristics; if the nutritional preferences of the parasites were known, an even greater separation might be possible.

There must be more to the story than mere topography. Based solely on microhabitat overlap, a pair such as *Himasthla-Lepocreadium* should theoretically be able to coexist; in practice this coexistence is never found. The histological picture in double combinations shows clearly how the parasite species in double infections affect each other.

In the first case history, *Zoögonus-Lepocreadium*, the parasites occur in adjacent mosaic clumps. In some infections complete coexistence occurs, with *Zoögonus* adjacent to *Lepocreadium* throughout, but in most instances there is a definite displacement of *Lepocreadium* to Area 4, anterior to the hepatopancreas, and to some extent into Area 3. *Zoögonus* tends to predominate in numbers, and to concentrate in the hepatopancreas and gonadal areas. In two instances, approximately 10% of the total parasites showed pycnotic nuclei; evidence suggests that they were *Lepocreadium*.

A second case history is that of the *Austrobilharzia-Cardiocephalus* infection. Here the story seems to be "good fences make good neighbors." The shistosome *Austrobilharzia* is markedly displaced towards the posterior apical end, predominating in the gonadal area. *Cardiocephalus* is more abundant anteriorly; only a few hepatopancreatic tubules remain intact. Even in the region of overlap of the two parasites, there is a segregation of the two species (Fig. 6).

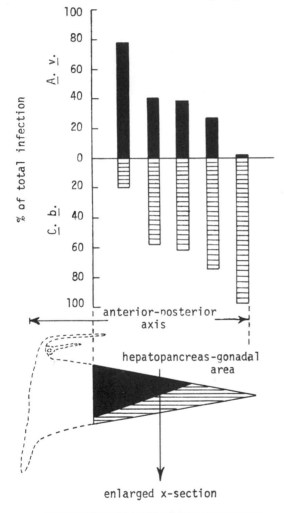

Figure 6. Double infection of larval *Cardiocephalus* and *Austrobilharzia* in *Nassarius obsoletus*. Schematic representation of distribution (above and middle) and photomicrograph of infected hepatopancreas, 30X (below). A=*Austrobilharzia* sporocyst, C=*Cardiocephalus* sporocyst.

A highly interesting case is the third example, *Himasthla-Austrobilharzia* (Fig. 7). *Austrobilharzia* is found almost exclusively in the hepatopancreas; only a few sporocysts are found in the apical gonadal area, which is the favored site in single infections. *Himasthla*, as in single infections, is found in the Leydig cell layer around the gut, and in the apical coil. A considerable number, however, are found in displaced positions such as the gill filaments, the kidney, and the connective tissues of the head. Furthermore, many are highly abnormal in appearance. Some parasites show signs of degeneration to such an extent that species identification is not possible. Others, particularly in the gills and kidney, seem to be encapsulated by the host.

Examples of microhabitat selection and displacement in

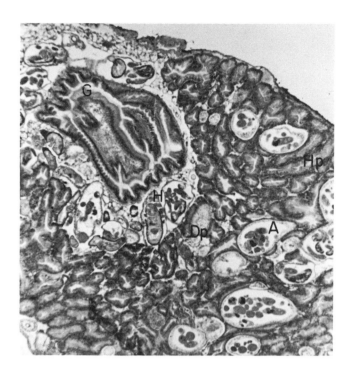

Figure 7. Photomicrograph of *Himasthla-Austrobilharzia* double infection of larval trematodes in *Nassarius obsoletus*. H=
Himasthla rediae; A=*Austrobilharzia* sporocyst; Hp=hepato-
pancreatic tubules; G=gut; Dp=degenerate parasite. 70X.

multiple helminth infections are well documented. Holmes (1961),
working with concurrent acanthocephalan and cestode parasites in the
rat gut, found that the choice location of both in single infections
was the anterior part of the intestine; in double infections the
acanthocephalan displaced the cestode parasite to a posterior position and little distributional overlap occurred. In the European
tortoise, *Testudo graeca*, Schad (1963) found ten or more species of
nematode (genus *Tachygonetria*). While longitudinal overlap occurred,
a combination of radial distribution differences and feeding preference differences greatly reduced interspecific competition. In
contrast, Lauckner (1971), working with sporocyst and metacercarial
generations of trematodes in two species of cockle (*Cardium* sp.)
found that where incidence and intensity of infection were greatest,
up to eleven of thirteen known species may parasitize the same host
specimen. Microhabitat selection occurred with some parasite species
but there was no evidence for interspecific exclusion. These data
suggest that displacement often occurs when helminth species overlap
in their primary habitat area.

    Another approach to the problem which sheds light on the com-
petition question is the controlled production of double infections
of trematode larvae in snails. In an extensive series of experiments, Lie and coworkers (Lie et al. 1968) were able to rear miracidia, and to control the infection with specific parasites in timed

sequence. By using a strain of thin-shelled, albino snails the progress of the infections and the interactions of multiple infections could be observed directly without sacrifice of the snails. Direct predation on the sporocysts and cercariae by rediae of echinostome species was seen in some cases (Lie 1966; Heyneman and Umathevy 1968). In addition to the direct antagonism, an indirect antagonism was noticed between sporocysts of a shistosome and strigeid species (Basch, Lie, and Heyneman 1969). Even more complex forms of antagonism involving hyperparasitism and snail host immunity have been reported (Basch 1970). Such observational techniques are extremely useful, but unfortunately are not applicable to the thick-walled, heavily pigmented marine snails.

A final clue to the population dynamics of double larval trematode infections is the biochemical alteration brought about by a parasite in the host's metabolic machinery. The possibility exists that modification of the metabolism by one parasite renders the tissues unsuitable for a competing species. Such an approach is intriguing, but has claimed scant attention in the literature. Alterations have been reported in the protein, carbohydrate, and lipid composition of a number of species of molluscan hosts which are parasitized by larval digenean trematodes (Reader 1971; Richards 1969; Wright 1966). Additional examples of parasite-induced modifications in host metabolism are reviewed by Cheng (1967). Several cases from our laboratory warrant special attention. Lunetta and Vernberg (1971) determined the normal concentration of a wide variety of fatty acids in the hepatopancreas of nonparasitized snails. With snails singly infected by *Lepocreadium, Zoögonus, Himasthla,* or *Austrobilharzia,* the host hepatopancreas was dissected free of parasites and analyzed similarly; striking quantitative changes were found (Fig. 8). Comparable evidence exists for enzyme systems. Figure 9 illustrates changes in cytochrome c oxidase concentration in the hepatopancreas of the same four infections and in noninfected snails, under warm- and cold-acclimated conditions (Vernberg and Vernberg 1968; Vernberg 1969). Whether these changes by one species are sufficient to act as controlling barriers to subsequent trematode infections is currently a matter of speculation. The suggestion has not been strongly enough stated or adequately documented, but definitely should be pursued further.

In summary, available information from our experiments and from the literature suggests that various kinds of interactions are possible between competing larval helminth species. These are as follows:

1. Complete coexistence without harmful interaction.
2. Preferential selection of different microhabitats within the gonad-digestive area, reducing competition.
3. Longitudinal or radial displacement in the tissue of choice, or displacement to other tissues.
4. Complete unilateral cannibalism, especially by echinostome rediae, with elimination or reduction of the other pair member.

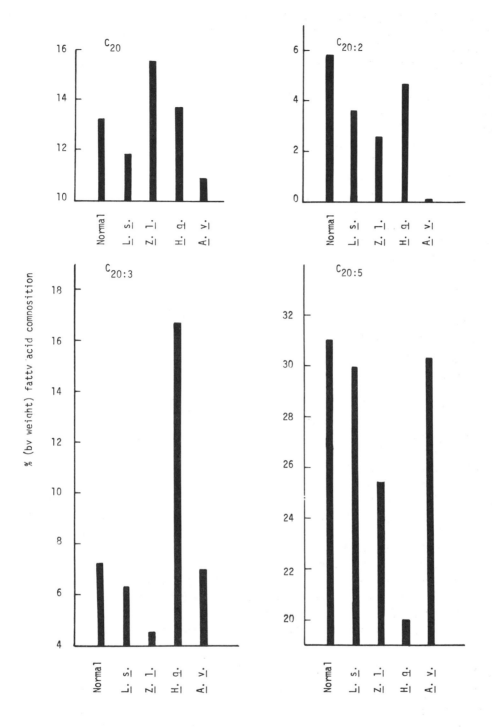

Figure 8. Relative percentage by weight of four C-20 fatty acids in the hepatopancreas of *Nassarius obsoletus*, either uninfected or singly infected with *Lepocreadium*, *Zoögonus*, *Himasthla*, or *Austrobilharzia*.

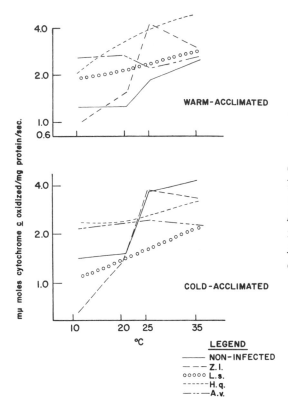

Figure 9.
Cytochrome c oxidase activity of hepatopancreas from warm-acclimated (above) and cold-acclimated (below) *Nassarius obsoletus*, either uninfected or singly infected with *Lepocreadium, Zoögonus, Himasthla,* or *Austrobilharzia.*

5. Indirect antagonism, resulting in the reduction or elimination of one species.
6. Hyperparasitism causing elimination or injury to one or both species.
7. Changed resistance of the host for one of the parasites, or changed environment in the host.

Thus, as with free-living animals, there is no single method of population regulation of helminths. Evidence from our data indicates that when two species of larval trematodes coexist without observable damage to each other, spatial adjustment on the part of each of them has occurred. In other combinations there is obvious antagonism between the two species, and it seems probable that eventually one will displace the other. The question of why certain species of larval trematodes are never found together in the same host is as yet unanswered. This could result from a direct antagonistic effect of one species on an invading one or perhaps one species has so metabolically altered the host tissue that it is impossible for the second species to survive.

ACKNOWLEDGMENTS

We wish to acknowledge the painstaking care of Mrs. Barabara Caldwell and Mrs. Elizabeth Clark in the preparation of the serial sections. Mrs. Dorothy Knight helped in the preparation of the manuscript and Miss Susan Ivester with the photomicrographs. The guidance of Dr. Bryan James during a visit to our laboratory, in the interpretation of the histological material was invaluable; many of his suggestions and ideas are incorporated into the manuscript. Dr. Thomas Cheng also offered helpful suggestions concerning the histological materials.

LITERATURE CITED

Ayala, F. J. 1972. Competition between species. Amer. Sci. 60: 348-57.

Basch, P. F. 1970. Relationships of some larval strigeids and echinostomes (Trematoda): Hyperparasitism, antagonism, and "immunity" in the snail host. Exptl. Parasitol. 27:193-216.

_____, Lie, K. J., and Heyneman, D. 1969. Antagonistic interaction between strigeid and schistosome sporocysts within a snail host. J. Parasitol. 55:753-58.

Boer, P. J. den, and Gradwell, G. R., eds. 1971. Dynamics of populations. Wageningen, The Netherlands: Centre for Ag. Publishing and Documentation.

Bourns, T. K. R. 1963. Larval trematodes parasitizing Lymnaea stagnalis appressa (Say) in Ontario with emphasis on multiple infections. Can. J. Zool. 41:937-41.

Cheng, T. C. 1962. The effects of parasitism by the larvae of Echnioparyphium Dietz (Trematoda: Echinostomatidae) on the structure of an glycogen deposition in the hepatopancreas of Helisoma trivolvis Say. Amer. Zool. 2:513.

_____. 1964. The biology of animal parasites. Philadelphia: W. B. Saunders Co.

_____. 1967. Marine molluscs as hosts for symbioses, with a review of known parasites of commercially important species. In Advances in Marine Biology, ed. F. S. Russell, Vol. 5. London and New York: Academic Press.

_____, and Cooperman, J. S. 1964. Studies on host-parasite relationships between larval trematodes and their hosts. V. The invasion of the reproductive system of Helisoma trivolvis by the sporocysts and cercariae of Glypthelmins pennsylvaniensis. Trans. Am. Microscop. Soc. 83:12-23.

_____, and Rifkin, E. 1970. Cellular reactions in marine molluscs in response to helminth parasitism. Symposium on diseases of fishes and shellfishes. Amer. Fish. Soc., Special Publication 5:443-95.

_____, and Snyder, R. W. 1962. Studies on host-parasite relationships between larval trematodes and their hosts. I. A review. II. The utilization of the host's glycogen by the intramolluscan larvae of Glypthelmins pennsylvaniensis Cheng and associated phenomena. Trans. Amer. Microscop. Soc. 81:209-28.

_____, and Yee, H. W. F. 1968. Histochemical demonstration of aminopeptidase activity associated with the intramolluscan stages of *Philophthalmus gralli* Mathis & Léger. <u>Parasitol.</u> 58:473-80.

Craig, L. C. 1972. The emergence patterns of two species of larval trematodes, *Himasthla quissetensis* and *Lepocreadium setiferoides* from their molluscan host, *Nassarius obsoleta*. Masters thesis, University of South Carolina.

Harger, J. R. E. 1972. Competitive coexistence among intertidal invertebrates. <u>Amer.</u> <u>Sci.</u> 60:600-07.

Heyneman, D., and Umathevy, T. 1968. Interaction of trematodes by predation within natural double infections in the host snail *Indoplanorbis exustus*. <u>Nature</u> 217:283-85.

Holmes, J. C. 1961. Effects of concurrent infections of *Hymenolepis diminuta* (Cestoda) and *Moniliformis dubius* (Acanthocephala). I. General effects and comparison with crowding. <u>J.</u> <u>Parasitol.</u> 47:209-16.

James, B. L. 1965. The effects of parasitism by larval Digenea on the digestive gland of the intertidal prosobranch, *Littorina saxatilis* (Olivi) subsp. *tenebrosa* (Montagu). <u>Parasitol.</u> 55: 93-115.

_____, and Bowers, E. A. 1967. Histochemical observations on the occurrence of carbohydrates, lipids, and enzymes in the daughter sporocyst of *Cercaria bucephalopsis haimaena* Lacaze-Duthiers, 1854 (Digenea: Bucephalidae). <u>Parasitol.</u> 57:79-86.

Lack, D. 1971. <u>Ecological</u> <u>isolation</u> <u>in</u> <u>birds.</u> Cambridge: Harvard University Press.

Lauckner, G. 1971. Zur Trematodenfauna der Herzmuscheln *Cardium edule* und *Cardium lamarcki*. <u>Helgoländer</u> <u>wiss.</u> <u>Meeresunters.</u> 22:377-400.

Levin, B. R. 1972. Coexistence of two asexual strains on a single resource. <u>Science</u> 175:1272-74.

Lie, K. J. 1966. Antagonistic interaction between *Schistosoma mansoni* and echinostome rediae in the snail *Australorbis glabratus*. <u>Nature</u> 211:1213-15.

_____, Basch, P. F., Heyneman, D., Beck, A. J., and Audy, J. R. 1968. Implications for trematode control of interspecific larval antagonism within snail hosts. <u>Trans.</u> <u>Roy.</u> <u>Soc.</u> <u>Trop.</u> <u>Med.</u> <u>Hyg.</u> 62:299-319.

Lunetta, J. E., and Vernberg, W. B. 1971. Fatty acid composition of parasitized and non-parasitized tissue of the mud-flat snail *Nassarius obsoleta* (Say). <u>Exptl.</u> <u>Parasitol.</u> 30:244-48.

Martin, W. E. 1955. Seasonal infections of the snail, *Cerithidae californica* Haldeman, with larval trematodes. In <u>Essays</u> <u>in</u> <u>the</u> <u>natural</u> <u>sciences,</u> pp. 203-10. Alan Hancock Foundation for Scientific Research. Los Angeles: Univ. Southern California Press.

Porter, C. A. 1970. The effects of parasitism by the trematode *Plagioporus virens* on the digestive gland of its snail host, *Flumenicola virens*. <u>Proc.</u> <u>Helminthol.</u> <u>Soc.</u> <u>Wash.</u> 37:39-44.

109

Reader, T. A. J. 1971. Histochemical observations on carbohydrates, lipids and enzymes in digenean parasites and host tissues of *Bithynia tentaculata*. Parasitol. 63:125-36.

Rees, W. J. 1936. The effects of parasitism by larval trematodes in the tissues of *Littorina littorea* (Linné). Proc. Zool. Soc. Lond. 2:357-68.

Richards, R. J. 1969. Qualitative and quantitative estimations of the free amino acids in the healthy and parasitized digestive gland and gonad of *Littorina saxatilis tenebrosa* (Mont.) and in the daughter sporocysts of *Microphallus pygmaeus* (Levinsen, 1881) and *Microphallus similis* (Jägerskiöld 1900) (Trematoda: Microphallidae). Comp. Biochem. Physiol. 31:655-65.

Robson, E. M., and Williams, I. C. 1971. Relationships of some species of Digenea with the marine prosobranch *Littorina littorea* (L.) II. The effect of larval Digenea on the reproductive biology of *L. littorea*. J. Helminth. 45:145-59.

Sastry, A. N. 1971. Effect of temperature on egg capsule deposition in the mud snail, *Nassarius obsoletus* (Say). Veliger 13:339-41.

Schad, G. A. 1963. Niche diversification in a parasitic species flock. Nature 198:404-06.

_____. 1966. Natural control of the abundance of parasitic helminths. Bull. Ind. Soc. Mal. Com. Dis. 3:18-27.

Valle, C., Pellegrino, J., and Alvarenga, N. 1971. Ritmo Circadiano de Emergencia de Cercarias (*Schistosoma mansoni* - *Biomphalaria glabrata*). Rev. Brasil. Biol. 31:53-63.

Vernberg, W. B. 1969. Adaptations of host and symbionts in the intertidal zone. Amer. Zool. 9:357-65.

_____, and Vernberg, F. J. 1968. Interrelationships between parasites and their hosts. IV. Cytochrome c oxidase thermal acclimation patterns in a larval trematode and its host. Exptl. Parasitol. 23:347-54.

_____, Vernberg, F. J., and Beckerdite, F. W., Jr. 1969. Larval trematodes: Double infections in the common mud-flat snail. Science 164:1287-88.

Wright, C. A. 1966. The pathogenesis of helminths in the Mollusca. Helminthol. Abstr. 35:207-224.

Evolutionary aspects of associations between crabs and sea anemones

D. M. Ross
Faculty of Science
Office of the Dean
The University of Alberta
Edmonton, Canada

To the unaided eye the phenomenon of cohabitation among sea anemones and hermit crabs is one of the most obvious examples of symbiosis. Both members of the pair are large, often colorful, and very distinctive because of their morphological disparity. Marine naturalists and enthusiasts have often commented on these pairs as objects of general interest and sometimes with picturesque descriptions, as when Philip Gosse (1860) spoke of the parasitic anemone as having "pitched its tent on the shell of the hermit."

Although these associations have been noticed and reported along with other curiosities and discoveries during the exploratory or collecting phase of marine biology, there are not many studies on the factors involved in these associations. Except for Faurot's (1910) long paper which dealt mainly with the cloak anemone, *Adamsia palliata*, and *Pagurus* (=*Eupagurus*) *prideauxi*, there is no major work from the classical period of zoology on the basic relationships between the two symbionts. About the same time, Brunelli (1910,1913), and later Brock (1927) at Naples described how *Dardanus arrosor* picked up their anemones. From time to time other authors wrote short papers and notes about crabs that carried anemones in various other places and how some of these, too, picked them up and placed them on their shells or claws (Cowles 1919). With such an obvious phenomenon one might assume that everything worth knowing was known and, indeed, this is the impression that one obtains from T. A. Stephenson's monograph on The British Sea Anemones (1928, 1935). Yet, as I discovered when I began to investigate this subject, little was known in any detail and this knowledge was dispersed throughout the zoological literature and not easily appraised.

A background for this paper has been provided by screening the film "Passengers or Partners?" beforehand. This film is essentially a review of what I have been able to find out by dabbling in this backwater for about ten years. What has emerged is a mixture of new observations tied in with older observations that have been reexamined and seen for the first time by one pair of eyes. The different associations can now be related to one another with some measure of consistency.

There are still many gaps in my personal experience that I should like to fill and there are many references to earlier work or records that should be followed up. Yet, having seen personally over thirty species-pairs, it seemed appropriate to open up the topic of the evolution of the habit of symbiosis in these animals for discussion at this meeting. There are four areas of relevant data.

(1) Taxonomic and systematic. There is much confusion in the literature on the exact identities of the taxa that are involved in these associations. Not only is one concerned with the taxonomy of crabs and anemones but one must also consider the gastropods that are sometimes hosts for anemones and whose shells are occupied by pagurids, and also the octopods, from whom the anemones seem to protect certain crabs (Ross 1971).

(2) Zoogeographical. The distribution of these associations on a world scale and locally can throw much light on the possible history of the associations. The general zoogeographical distribution is now known fairly accurately but the local distributions much less well.

(3) Behavioral and physiological. The activities and adaptations that each symbiont shows for establishing the association have now been recorded in many cases. These have an obvious application to the problem of the evolutionary origin of these species-pairs.

(4) Ecological. It is important to know where these associations fit into the ecological scheme in various marine communities but exact data on this matter are scanty. Thus, discussions on the selective advantages that might accrue from living in these associations are still somewhat speculative.

TAXONOMIC AND SYSTEMATIC

Taxonomic details are murky because of many uncertainties about the identifications in the older literature and the various upheavals that have taken place in naming and classifying both hermit crabs and sea anemones. The species of crabs and anemones that live together fall into a relatively restricted number of taxonomic groups. Most of the crabs are in the Anomura and, according to an older taxonomic scheme, all these would have been in the Family Paguridae. A newer scheme (MacDonald, Pike, and Williamson 1957) erected two superfamilie

(Coenobitoidea and Paguroidea). This arrangement separated hermits of the genus *Pagurus* in the Paguroidea from another family, the Diogeneidae, in which most of the hermits involved in these associations with anemones are actually found. Some brachyuran crabs also carry anemones, e.g., a few majids and calappids.

There is no pattern in the symbiotic associations at the family level in these basically cosmopolitan families. Within the families it is usually a case of particular and sometimes scattered species showing the habit, except for the genus *Dardanus* in the Family Diogeneidae. Almost all species of *Dardanus* carry anemones on their shells. I have personally collected ten species of *Dardanus* and only one lacked an anemone associate and failed to pick up anemones found in such associations and place them on its shell. Among the large genus *Pagurus*, only *P. prideauxi* is consistently associated with an anemone, i.e., *Adamsia palliata*. The other species are occasional symbionts or symbionts within limited parts of their ranges. The same is true of some other genera, e.g., *Anapagurus* and *Paguristes*.

Some brachyurans are seldom found without anemones, e.g., the Caribbean decorator crab, *Stenocionops furcata*. But the records need to be extended. Unfortunately, single records only are available in many cases and, if the collector happened to be interested in anemones, the crab was usually discarded and vice versa. An example of the discoveries and rediscoveries still to be made is seen in Provenzano's (1971) collection of *Munidopagurus macrocheles*, which had not been seen since 1880, showing that it has a fairly wide distribution in the Caribbean and to the north. This hermit is always found with a single unidentified anemone (Fig. 1) living like the cloak anemone of European Atlantic Coasts and the Mediterranean, but unlike *P. prideauxi*, *M. macrocheles* has dispensed altogether with a shell. It does not even retain the small shell or shell fragments which seem to be necessary for *P. prideauxi* (author's unpublished observation).

Another apparently obligate association is the so-called "boxer crab" found in the Central and West Pacific, *Lybia tesselata* (=*Melia tesselata*), which carries anemones on its claws. This association has not been studied since Duerden (1905) wrote it up early in the century. I have not seen it myself.

Turning now to the anemones, we find that they are mostly restricted to a single family, the Hormathiidae. Within that family, three genera, *Calliactis*, *Paracalliactis*, and *Adamsia*, are so rarely found not living on crabs that they must be regarded essentially as obligate symbionts which have evolved in this direction. Outside these genera there are a few well-documented examples of Hormathiidae living with crabs, though some live with gastropods. Indeed, the European *Hormathia digitata*, from which the family gets its name, is almost always found on living gastropods.

In other families of anemones, a few examples of associations with crabs exist. In the Actinostolidae, *Antholoba achates* is frequently found living on crabs on the western coast of South America,

Figure 1. *Munidopagurus macrocheles* with unidentified sea anemone
covering abdomen, as collected in Western Atlantic.
(Photo courtesy Dr. Anthony Provenzano, Jr.)

and in southern Japan, the small *Paranthus sociatus* is apparently an
obligate symbiont on two small gastropods. In the Family Actiniidae,
two species found living on crabs and for which only single records
exist have been assigned to the genera *Isadamsia* (Carlgren 1928) and
Isocradactis (Stuckey 1909) in the Indian Ocean and in New Zealand,
respectively. I have studied four other species not yet adequately
identified, that have been found living only in associations. Preliminary identifications in some cases have attached these to genera
in other families, e.g., *Sagartiomorphe* assigned to the Family Sagar-
tiomorphidae (Carlgren 1949).

These isolated records and unidentified species are possibly on the periphery of the main phenomenon of symbiosis in these animals. It happens that the anemones and the crabs which are most frequently involved in these association are well known and their systematic positions are fairly well established. It is quite conceivable that further study will lead to extensive revision of the assignments of the lesser known species to various genera and even to families. The taxonomic picture suggests, therefore, that if we knew something about the history and evolutionary relationships of the genus *Dardanus* and some species of *Pagurus* among the hermit crabs, and about *Calliactis*, *Paracalliactis* and *Adamsia* in the Family Hormathiidae among the anemones, it might be possible to make meaningful comments about the evolution of the habit of symbiosis among these creatures.

ZOOGEOGRAPHICAL

The zoogeography of crab-anemone associations seems to me to be very significant. They are almost everywhere confined to warmer seas and at depths less than a few hundred meters. On the map of the world (Fig. 2) it is essentially a circumtropical distribution with some warm temperate extensions. There are some regions, however,

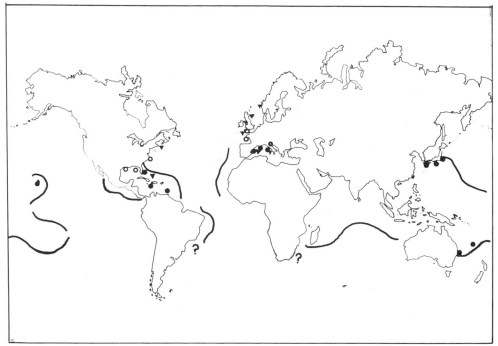

Figure 2. Geographical distribution of some anemone-crab associations. ●=*Calliactis-Dardanus* associations (author's collections only); ○=*Calliactis* spp. in associations outside
range of *Calliactis-Dardanus* associations; ▼=Cloak-anem-
ones, e.g., *Adamsia* spp.; ——=Impression of boundaries of
Calliactis-Dardanus associations; ?=Areas of uncertainty.

e.g., the West Tropical Coasts of the Americas, where the associations are neither extensive nor numerous. The best known exception
to this tropical picture is the association between *Adamsia palliata*
and *Pagurus prideauxi*, which extends from the Mediterranean up the
Atlantic Coast of Western Europe as far as Norway. Also in New Zealand there are associations of anemones, whose identities Cadet Hand*
refers to *Calliactis conchicola* and *Paracalliactis rosea*, that extend
outside the typically tropical and subtropical range.

Carlgren (1949) lists fourteen species of *Calliactis*. Charles
Cutress** who has made a study of these is of the opinion that some
of these species are dubious and certainly some of those listed as
distinct species in the Indian and Pacific Oceans should be assigned
to *Calliactis polypus*. In his general description of the genus
Calliactis, Carlgren (1949) stated that it is "often commensal with
hermit crabs." In fact, the three species that are well known, *C.
parasitica, C. polypus* and *C. tricolor*, are almost never found
except as symbionts. Moreover, the vast majority of *Calliactis*, as
individuals and as species, are found on different species of one
genus, *Dardanus*. The only important exceptions are: 1) *C. parasitica*
being carried by *P. bernhardus* on the coast of Western Europe; 2)
C. tricolor being carried by *P. pollicaris* and *P. impressus* and by
the majid, *Stenocionops furcata*, and the calappid, *Hepatus epheliti-
cus* in the West Indies and on the southeastern coasts of the United
States and in the Gulf of Mexico.

The boundaries of the associations and the participating species
show some significant points. Thus, in Japan, the association and
the participating species, *Calliactic polypus* and *Dardanus* spp., seem
to disappear together somewhere just north of Tokyo. The same situation occurs in the region of Sydney, Australia, where *C. polypus* and
Dardanus also fade out together. This shows that in these locations
the two species live in a state of strong mutual dependence and
possibly also that they occupy these territories together as well-
established symbionts.

A special feature of the distribution of anemone symbionts is
the relative infrequency of their associations with living gastropods. Especially significant is the case of *C. parasitica* which is
found on *P. bernhardus* in the English Channel in immense numbers.
Yet, although *P. bernhardus* inhabits *Buccinum* shells, the anemone
is not found on the living *Buccinum*. A different situation exists
in the Adriatic Sea. Here *C. parasitica* is found in the upper Adriatic near Rovinje and Venice, outside the range of *D. arrosor*, its common host in the rest of the Adriatic.*** Apparently, *D. arrosor* cannot live in the waters of the upper Adriatic, yet *C. parasitica* is no
not so restricted and in these waters it uses living gastropods as
alternative hosts. It happens that *C. parasitica* displays a behavior
pattern which enables it to live in this way (Ross and Sutton 1961).

---

*Cadet Hand: personal communication.
**Charles Cutress: personal communication.
***K. Lorenz, D. Mainardi, and A. Rossi: personal communications.

Other species of *Calliactis*, e.g., *C. polypus* (Ross 1970), lack this capacity and they are confined to living on active crabs.

Other anemone-gastropod partnerships turn up in various places. The anemones in New Zealand most commonly found living with hermit and spider crabs are also found occasionally on living gastropods (author's unpublished observation). In Japan I was able to study one unidentified species that was found living only on a single species of gastropod, *Siphonalia filosa*. I have also examined *Hormathia digitata* found living on *Neptunea antiqua* off the coasts of Sweden and Norway. Other anemones have been reported as occurring on gastropods in more northern locations but information on distribution and identities of hosts is scanty. Even from these few examples, however, an obvious conclusion emerges that gastropod-anemone partnerships are almost a converse of crab-anemone associations. They appear on the periphery of the associations between crabs and anemones and are particularly associated with boreal and cold temperature conditions.

BEHAVIORAL AND PHYSIOLOGICAL

Three patterns of behavior are involved in the establishment of associations between crabs and anemones. (1) The active transfer of anemones to shells occupied by hermit crabs or to living mollusks. There are two variations of this: (a) the complete transfer of firmly attached anemones to shells from surfaces that are not shell, e.g., stones, glass, plastic, by detaching the foot and somersaulting across the gap to resettle on the shell; and (b) the clinging to shells and settling of anemones that have already been detached (Ross and Sutton 1961). (2) The active detachment of anemones by crabs using their walking legs and claws to relax and ease the anemone off the surface to place it on the shell (Brunelli 1913; Brock 1927; Cutress, Ross, and Sutton 1970). (3) The anemone's relaxation and detachment in response to the stimulation of an active crab or to stimuli resembling the crab's activities set up by various substitutes (Ross and Sutton 1970).

It is instructive to discuss the occurrence of these behavior patterns in the different genera and species of crab-anemone symbionts that have been examined.

Response to Shells

The transfer of attached anemones to shells seems to be well-developed only in *Calliactis parasitica*. *C. tricolor* shows the same response but perhaps not when settled for long periods. It has never been observed in *C. polypus*. Anemones living on gastropods, e.g., *Hormathia digitata*, have never been available in the numbers or in the conditions necessary for a satisfactory testing of their ability to transfer from other surfaces to shells. Presumably this must occur at some stage in the life history but good observations are lacking.

Anemones living in associations with crabs all show capacities to cling to shells and to climb onto shells when already detached but this behavior pattern seems to be more pronounced in *C. parasitica* than in the other species of *Calliactis*. *Paracalliactis japonica* shows a special variation; even when detached, this anemone required preliminary stimulation by a crab before clinging and settling on shells (author's unpublished observation). I have much information demonstrating that the actual method by which each *Adamsia* species of anemone clings and settles on shells is highly characteristic of the species; a wide range of techniques is employed in different cases.

Activity by Crabs

Active crabs include all species of *Dardanus* which carry anemones. Others are the Caribbean decorator crab, *Stenocionops furcata*, and some other members of the Diogeneidae which have been available, e.g., *Petrochirus diogenes*. The only consistently active species of *Pagurus* is *P. prideauxi*, which is an obligate symbiont with *palliata*. When deprived of its *Adamsia*, *P. prideauxi* attempts to pick up another as soon as it make contact. Some species of *Pagurus* that carry *Calliactis*, *P. bernhardus* (Ross 1960) and *P. pollicaris*[*], have never displayed any overt activity towards their symbionts although in certain places they are almost always found together. The only exception is *P. impressus* which displays an active behavior pattern toward *C. tricolor*[*].

Relaxation and Detachment in
Response to the Crab's Manipulations

This behavior pattern is seen in all anemones that live as commensals with crabs. English populations of *C. parasitica* not normally exposed to this activity show the same response. The main events in this process are the state of relaxation produced in the anemone in response to gentle repeated mechanical stimuli applied to the column. This response also includes the dissolution of some adhesive mucus or cement that anchors an attached anemone to its location. An extension of this behavior pattern is the adhesiveness of the tentacles and pedal discs of any anemone detached in this way. Although this adhesiveness appears to be particularly strong toward shells, it is displayed toward other surfaces as well.

It was not possible to test the responses of all anemone symbionts, including those living on gastropods, either to the manipulations of crabs or to the various substitutes for crabs that have

---

[*]Mariscal: personal communication.

been employed to bring about relaxation and detachment (Ross and Sutton 1970). In my experience substitutes are not equally effective even in the three different species of *Calliactis*. Yet conditions seem to be very important and sometimes these tests were done in places where conditions were far from ideal. This is one case in which negative results must be treated with reserve.

The physiological analysis of these behavior patterns has not gone very far. In crabs displaying activities toward their anemones, some innate releasing mechanism is triggered by contact with the anemone. But one has only to watch the performance to see how far we are from understanding such a complex behavioral pattern in decapod crustacea, e.g., the pre-testing of the carapace in *Stenocionops furcata* (Cutress, Ross, and Sutton 1970).

In the anemones, the physiological analysis of the behavior patterns is even more of a problem. Their nervous systems do not seem to possess the equipment to enable them to do what they do. The response to shells is a very complex maneuver but the anatomical and physiological basis for it must exist in the nervous pathways, and it is remarkable that they switch effectors on and off in response to the flow of excitation from the shell through the tentacles to the foot. Nor is the shell always equally effective. The response to a shell depends in part on how long and how firmly the foot has been settled on the other surface from which it is transferring. If it has recently been occupied by an anemone it ceases to be attractive (author's unpublished observation).

Using low frequency electrical stimuli, it is possible to produce relaxation and detachment in different species of *Calliactis* (Ross and Sutton 1968; Ross 1970; MacFarlane 1969). These stimuli apparently mimic the gentle tactile stimuli applied by the crab and by substitutes in human hands, such as pipe cleaners and brushes. It is not yet known whether low frequency electrical stimuli can bring about detachment in all anemones that live in associations with crabs. Results so far indicate that comparable results do not occur in anemones that do not live in associations with crabs but more studies are required under standardized conditions. The behavior patterns in which active crabs are involved are the most widespread and complete. The crab's performance and the anemone's response to this performance and to the shell are all beautifully integrated into mutual programs of both species in producing rapid and successful transfers. Other less perfect joint behavior patterns and one-sided transfers by anemones raise questions of relationships that can only be examined when more data are available.

ECOLOGICAL

The ecological and adaptational aspects of the crab-anemone symbioses are crucial to an understanding of their possible evolution. Unfortunately, this is one of the areas of maximum ignorance.

From collections obtained by dredging and from laboratory experiments one can only make inferences about the precise relationships that exist in waters beyond the shore. Diving has not given much information on this topic as yet but it should help greatly in the future.

Three types of associations are clearly distinguished from one another: (1) associations in which the anemone is apparently a passenger only; (2) associations in which the anemone protects the crab by forming a cloak and substituting for the shell; and (3) associations between anemones and active crabs in which the anemone probably protects the crab from predators.

On inactive crabs (some species of *Pagurus* and *Paguristes*) and gastropods, one must assume that the anemone as the active member in establishing the association is the main or sole beneficiary. A plausible and attractive explanation for this habit assumes that an anemone which is being transported will have a wider choice of food and may escape from some unsatisfactory conditions which might affect it adversely in a fixed location. Such advantages are difficult to demonstrate.

Where the anemone serves as a cloak, as in *Adamsia palliata*, it provides the hermit crab with a means of avoiding that critical and sometimes fatal necessity of replacing the shell as it grows. *Munidopagurus macrocheles* has gone even further in this direction by dispensing with the shell altogether (Provenzano 1971). The problematical *Isadamsia* described by Carlgren (1928) and *Adamsia sociabilis* described by Verrill (1883) from North Atlantic collections indicate only a modest tendency for the evolution of the cloak anemone type of partnership on a world scale.

The associations of anemones and active crabs of the *Calliactis-Dardanus* type are more widespread and successful. At last there is a good behavioral hypothesis to explain the intense activity displayed by crabs in placing *Calliactis* on their shells, i.e., protection from *Octopus vulgaris* (Ross 1971). The laboratory experiments could not have been more convincing and the distribution of the animals fits almost perfectly (Fig. 2). These laboratory observations, however, need now to be followed up by field studies to see whether they can be confirmed by observations in natural habitats.

Interesting possibilities for further behavioral studies arise from these observations with *Octopus*. There are opportunities here for testing memory retention and discrimination in *Octopus*, using natural rather than artificial experimental situations. Boycott (1954) did some preliminary work along this line. The behavior of *Dardanus* in these situations also opens up another avenue in that *Dardanus* with anemones behave differently (described as "aggressive" or "dominant") from those without anemones (author's unpublished observations). In this connection, one notes the work of Mainardi and Rossi (1969), who adopted a converse approach and described how dominant *Dardanus* collect and pick up anemones while subordinate *Dardanus* submit and avoid the competition for anemones.

Another feature of the relationship between crabs and anemones, which is of profound ecological significance but largely unexplored, is the origin of the association in the lives of the crabs and the anemones in their natural habitats. In most cases the associations are known only as associations between adults. At what stages in their lives the individuals of the species-pair come together is still largely unknown. It is remarkable how seldom one discovers *Calliactis* of subadult size living on shells in these associations. In my experience, the only concentration of small *Calliactis* ever observed has been on the legs of the Caribbean decorator crab, *S. furcata*. The situation is different with *Adamsia palliata* which becomes associated with young *P. prideauxi* as a small symmetrical anemone and only develops later into the cloak form (Faurot 1910).

EVOLUTIONARY CONSIDERATIONS

Two possibilities arise when considering how this fascinating array of species-pairs might have emerged in the course of evolution. Did anemones first become associated with crabs by attaching themselves to shells either of gastropods or of hermit crabs? Or did anemones first become associated with crabs when hermits began to detach them and place them on their shells? Both possibilities have points in their favor and both are confounded by certain of the observed facts as described above.

For general reasons, and because only one adaptational step is involved, I prefer the notion that such associations probably began by a tendency among certain anemones to respond to shells and settle on them for the transport so obtained. Within this type of explanation one might argue for a priority in settling on gastropods first by noting that the response is based on the intensity of the shell factor and that factor is stronger in shells of living gastropods than in discarded gastropods used by hermits. This is true of *Calliactis parasitica* (Ross and Sutton 1961). One could argue that from having established associations with gastropods, this tendency could easily be transferred to the gastropod shells of certain hermit crabs which provided an even more mobile existence. By a further simple step, hermit crabs by virtue of the protection afforded by such passengers, possibly against *Octopus*, might have developed a behavior pattern toward tolerating and eventually toward the acquisition of such passengers, with a whole series of consequences on the mutual behavior patterns of both crabs and anemones as evolution progressed.

There are difficulties in this type of explanation. For one thing, the number of anemones living on gastropods is relatively insignificant compared with what one might expect if living on gastropods had definite advantages. Second, the number of anemones living as passengers on inactive hermit crabs is also very small compared with what might be expected if this were a way of life with considerable selective advantages. Third, *Adamsia palliata,* which lives with *Pagurus prideauxi* in the most intimate of all associations, shows no response to shells when detached (author's unpublished observation).

The alternative explanation is that the crab-anemone association began by crabs possibly using anemones to decorate their shells, as some decorator crabs and hermit crabs do with other organisms such as algae and sponges, presumably for concealment (Schöne 1968). It is conceivable that such a habit in crabs of the Family Diogeneidae conferred a selective advantage on the crabs and became perpetuated in the course of evolution. The processes by which such changes became perpetuated are generally referred to as natural selection but it must be admitted that in the field of behavior there are difficulties in relating behavioral changes to mechanisms of heredity as we know them. In any case this would explain how the associations happened to be concentrated in a few genera and species on both sides and particularly in *Dardanus* and *Calliactis*, since the latter provided protection for the crabs.

The extension of *Calliactis* as a symbiont of other nonactive species of the genus *Pagurus*, or to *Murex* in the upper Adriatic, would be explained as range overshoots of no great significance. They would reflect the fact that in adapting to the hermit crab association, a response to shells had been independently acquired by the anemone. The occurrence of other anemones on gastropods would then be regarded as independent evolutionary events of purely local significance. That this habit is not necessarily linked with the associations of some anemones with hermit crabs is indicated by the active behavior patterns by which some other actinians settle on molluscan shells, e.g., *Stomphia coccinea* on the lamellibranch *Modiolus modiolus* (Ross 1965).

The second explanation has the general difficulty that it requires crabs and anemones to have developed cooperating behavior patterns simultaneously. There are specific problems also. It is a striking fact that *Dardanus* can pick up only very few species of anemones and attempts to pick up nonsymbionts are generally quickly abandoned (author's unpublished observation). This suggests some preadaptation by the anemones in such partnerships, which would have been provided by the response of *Calliactis* to molluscan shells.

When one surveys the whole field it becomes obvious that these associations probably evolved independently on a good many different occasions. Thus, the cloak anemone association probably evolved before any of the other existing associations if one is to judge by the morphological changes that have taken place in *Adamsia palliata*. At the same time, there is no reason to believe that this association could have been the precursor or that it could have triggered off in any way the events that led to the establishement of the more widespread associations that now exist between various crabs and species of *Calliactis*.

The occurrence of other isolated associations, sometimes involving members of other actinian families, suggests very strongly that many of the examples that we see today could have arisen independently in different places at different times. Our main problem here is to see if there is an evolutionary pattern in the associations that now comprise the main populations of symbiotic sea anemones and crabs. This means essentially the *Calliactis-Dardanus* associations. In numbers and in range these eclipse all other members of the Family Diogeneidae among the crabs and other genera in the Family Hormathiidae among the anemones.

If the essential element in the association were the activity of the crab, it would seem unlikely that the anemones involved in these associations would ever have developed an independent response to shells enabling them to transfer by themselves. The fact that some anemones do transfer unaided argues that the associations began by anemones transferring to shells. It would be a plausible step arising from the protection given to the crabs by the anemones for crabs with anemones to have a selective advantage, and in subsequent evolution for the tendency to acquire the anemones to become part of the fixed behavioral repertoire of crabs living in places where such a selective advantage was enjoyed. The advantage of this interpretation is that it requires as an additional step on the part of the crab only that it assist the transfer by remaining with the anemone and helping it to do something it can do by itself. The subsequent evolution of the behavior pattern to the point where the activity of the crab seems to be the primary event and the shell response begins to decline seems to be a natural development in that a behavior pattern that is little used might be expected to disappear.

Relating these facts to the situation that exists at the present time in the various associations and populations of anemones and crabs living symbiotically, one can see that the Caribbean and the Indo-Pacific associations have moved further in the direction of independence from the anemone's response to the shell. This would imply that these species were more recently evolved but, unfortunately, there seems to be no way of testing this point. At the same time, the selective advantage in carrying *Calliactis* seems to have led some other species (e.g., *Stenocionops furcata*) to adopt the habit; perhaps this is one of the more recent evolutionary developments in these associations.

If the symbiosis between crabs and anemones began as associations between gastropods and anemones, two obvious questions arise. Why are most anemone symbionts now found on hermit crabs? Why are the main populations of *Calliactis* confined largely to crabs and not shared also with gastropods in the same location? The answers probably lie with the activity of crabs, especially of the various species of *Dardanus*. Once these crabs became active participants in establishing the partnerships, the balance was swung heavily in favor of crab-anemone associations over gastropod-anemone associations in any location where both associations might exist side by side. This would be bound to happen because the gastropod is merely a passive carrier whereas the crab is active enough even to remove the anemones from the gastropods should the opportunity arise. This explains why it is only when active crabs fade out, as in the upper reaches of the Adriatic, that significant numbers of *Calliactis* appear living on gastropods. This is exactly what one would expect if an original gastropod-*Calliactis* symbiosis had been appropriated in evolution for the selective advantage that it conferred on crabs that live in gastropod shells.

ACKNOWLEDGMENTS

The author acknowledges with thanks the permission given by Dr. Anthony Provenzano, Jr. and the University of Miami Press to reprint the photographs in Figure 1, originally published in Bull. Mar. Sci. Gulf & Carib., Vol. 21 (1971).

LITERATURE CITED

Boycott, B. B. 1954. Learning in *Octopus vulgaris* and other Cephalopods. Pubbl. Staz. Zool. Napoli 25:67–93.

Brock, F. 1927. Das Verhalten des Einsiedlerkrebses *Pagurus arrosor* Herbst während des Aufsuchens. Ablösens und Aufpflanzens seiner Seerose *Sagartia parasitica* Gosse. Wilhelm Roux Arch. Entwicklungsmech. Organismen 112:205–38.

Brunelli, G. 1910. Osservasioni ed esperienze sulla simbiosi dei Paguridi e delle Attinie. Atti Accad. Naz. Lincei Rend. Cl. Sci. Fis. Mat. Natur. 19:77–82.

_____. 1913. Richerche etologiche. Osservazioni ed esperienze sulla simbiosi dei Paguridi e delle Attinie. Zool. Jahrb. Abt. Allg. Zool. Physiol. Tiere 34:1–26.

Carlgren, O. 1928. Zur Symbiose zwischen Aktinien und Paguriden. Z. Morph. u. Ökol. 12:165–73.

_____. 1949. A survey of the Ptychodactiaria, Corallimorpharia and Actiniaria. Kgl. Svenska Vetensk-Akad. Handl. 1 (Ser. 4):1–121.

Cowles, R. P. 1919. The habits of tropical Crustacea. III. Habits and reactions of hermit crabs associated with sea anemones. Philippine J. Sci. 15:81–90.

Cutress, C., Ross, D. M., and Sutton, L. 1970. The association of *Calliactis tricolor* with its pagurid, calappid, and majid partners in the Caribbean. Can. J. Zool. 48:371–76.

Duerden, J. C. 1905 On the habits and reactions of crabs. Proc. Zool. Soc. London 2:494–511.

Faurot, L. 1910. Etudé sur les associations entre les Pagures et les Actinies; *Eupagurus prideauxi* Heller et *Adamsia palliata* Forbes, *Pagurus striatus* Latreille et *Sagartia parasitica* Gosse. Arch. Zool. Exp. Gen. 5:421–86.

Gosse, P. H. 1860. Actinologica Britannica. A history of the British sea-anemones and corals. London: Van Voorst.

MacDonald, J. D., Pike, R. B., and Williamson, D. I. 1957. Larvae of the British species of *Diogenes, Pagurus, Anapagurus* and *Lithodes* (Crustacea, Decapoda). Proc. Zool. Soc. London 128:209–57.

MacFarlane, I. D. 1969. Co-ordination of pedal-disk detachment in the sea anemone *Calliactus parasitica*. J. Exp. Biol. 51:387–96.

Mainardi, D., and Rossi, A. C. 1969. Relations between social status and activity towards the sea anemone *Calliactis parasitica* in the hermit crab *Dardanus arrosor*. Atti Accad. Rend. Naz. Lincei Cl. Fis. Mat. Natur. 8th Ser. 47:16–21.

Provenzano, A. J., Jr. 1971. Rediscovery of *Munidopagurus macrocheles* (A. Milne-Edwards, 1880) (Crustacea, Decapoda, Paguridae) with a description of the first zoeal stage. Bull. Mar. Sci. 21:256–66.

✓ Ross, D. M. 1960. The association between the hermit crab *Eupagurus bernhardus* (L.) and the sea anemone *Calliactis parasitica* (Couch). Proc. Zool. Soc. London 134:43–57.

_____. 1965. Preferential settling of the sea anemone *Stomphia coccinea* on the mussel *Modiolus modiolus*. Science 148:527–28.

_____. 1970. The commensal association of *Calliactis polypus* and the hermit crab *Dardanus gemmatus* in Hawaii. Can. J. Zool. 48:351–57.

_____. 1971. Protection of hermit crabs (*Dardanus* spp.) from octopus by commensal sea anemones (*Calliactis* spp.). Nature 230:401–02.

Ross, D. M., and Sutton, L. 1961. The response of the sea anemone *Calliactis parasitica* to shells of the hermit crab *Pagurus bernhardus*. Proc. Roy Soc. B155:266–81.

_____, and _____. 1968. Detachment of sea anemones by commensal hermit crabs and by mechanical and electrical stimuli. Nature 217:380–81.

_____, and _____. 1970. The detachment of the commensal sea anemones, *Calliactis polypus* and *C. tricolor* by mechanical and electrical stimulation. Z. vergl. Physiol. 67:102–19.

Schöne, H. 1968. Agonistic and sexual display in aquatic and semi-terrestrial brachyuran crabs. Amer. Zool. 8:641–54.

Stephenson, T. A. 1928 and 1935. The British sea anemones, Vols. 1 and 2. London: Ray Society.

Stuckey, F. G. A. 1909. A review of New Zealand Actiniaria. Trans. N. Z. Inst. 41:374–98.

Verrill, A. E. 1883. Report on the Anthozoa and on some additional species dredged by the "Blake" in 1877–79 and by the U. S. Fish Commission Steamer "Fish Hawk" in 1880–82. Bull. Mus. Comp. Zool. Harvard Coll. 11(1):1–72.

Symbioses in the Turbellaria and their implications in studies on the evolution of parasitism

J. B. Jennings
Department of Pure and Applied Zoology
University of Leeds
Leeds, England

It is generally believed that the Turbellaria are closer to the ancestral platyhelminth stock than are either the Trematoda or the Cestoda, and the widespread acceptance of this idea is shown by the fact that it is implicit in most theories concerning the evolutionary relationships of the lower Metazoa. While these differ from each other in many important details the majority agree in the final proposition of a common turbellarian like ancestor for the Platyhelminthes, and at least two of the major theories suggest an acoeloid form for this early metazoan.

The main morphological and and physiological differences between the Turbellaria, which are typically free-living predators, and the parasitic flatworms which constitute the majority of the phylum, can be correlated with the different modes of life. These differences, however, are in some instances very much a matter of degree, and often can be regarded simply as different levels of elaboration of similar, sometimes homologous, structures and processes. From this viewpoint the phylum is remarkably uniform and many characteristics of the Trematoda and the Cestoda are seen to be foreshadowed in the Turbellaria. Thus, apart from the obvious common features which have resulted in the recognition of the Platyhelminthes as a valid phylum, there are many others. For example, devices for adhesion and attachment appear in the Turbellaria as mucoid secretions, adhesive papillae, and simple suckers, and reach a climax of development in the complex suckers, supplemented by hooks and spines, of the parasitic classes. In other instances, though, there is very little elabora-

tion on the basic turbellarian situation, as in the case of the reproductive system. As Hyman (1951) points out, this system in free-
living flatworms is virtually as complicated as in the parasitic
ones due, no doubt, to the almost universal occurrence of hermaphro-
ditism. Life cycles, in contrast, are much more complex in the
parasitic flatworms and have apparently been evolved, in part at
least, as solutions to the problems presented by the mode of life.

Physiologically, too, most variations from the turbellarian norm
can be related to different life styles. In nutritional physiology,
for example, the situation in the Monogenea and Digenea has its origins in the clear-cut patterns characteristic of the Turbellaria, but
is somewhat obscured by supplementary modifications in digestive
physiology concerned with adaptation to the different diets made
available by the parasitic habit (Jennings 1968a). Similarly, respiratory physiology has been modified from that characteristic of the
free-living types (von Brand 1966).

In addition to all these morphological and physiological characteristics which are present in varying degrees throughout the phylum
there is, of course, one very obvious feature which is implicit in
the description of the Trematoda and the Cestoda as "parasitic."
This, simply, is the inherent tendency of flatworms to form associations, or symbioses, with other organisms and since parasitism is the
dominant life style of the phylum, this tendency is surely as characteristic of the taxon as is bilateral symmetry, the absence of a
coelom and anus, the dorsoventral flattening of the body. It occurs
in the Turbellaria to a much smaller extent than in the other classes
but, paradoxically, the turbellarians have evolved a much wider range
of types of symbiotic relationships. Approximately twenty-seven turbellarian families have members which enter into symbioses with higher
organisms. While eleven of these fall within the order Rhabdocoela,
the remainder are representative of all the other major subdivisions
of the class, except for the freshwater and terrestrial triclads
which are entirely free-living. This scattered distribution of symbioses throughout the Turbellaria is further evidence that the tendency to form such associations is a basic platyhelminth feature.

Since it has been established that throughout the Platyhelminthes certain characteristic traits occur which it is believed must
have their origins in the properties of some common turbellarian or
turbellarian like ancestor, an examination of these traits in the
modern Turbellaria is obviously worthwhile, in the hope of seeing
them manifested in something like their original form unmodified by
the special demands of a parasitic existence. I propose to examine
the various symbioses entered into by the Turbellaria, to assess the
nature and extent of the host-symbiont interaction, and to determine
to what extent the symbioses demonstrate pathways which could have
been followed by the Trematoda and Cestoda during their evolution
into their present form. In this context only those associations in
which other organisms act as hosts for turbellarians will be considered; the Turbellaria themselves act as hosts for a variety of
unicellular algae, flagellate, ciliate, and gregarine protozoa, and

occasionally nematodes and larval trematodes, but these associations are beyond the scope of the present account.

BRIEF SYSTEMATIC REVIEW OF SYMBIOTIC TURBELLARIA

A detailed systematic survey of the families, genera, and species of symbiotic Turbellaria, together with an extensive biblio-graphy, has been given elsewhere (Jennings 1971). My emphasis will be upon the types of associations, rather than on the systematics of the turbellarian symbionts, but a brief general review of the latter down to the level of families will provide a useful background orien-tation.

Tables 1 and 2 summarize those turbellarian families which con-tain symbiotic species, and also show the types of host organisms concerned in the associations. It can be seen that the most common hosts in turbellarian symbioses are echinoderms (Asteroidea, Ophiu-roidea, Echinoidea, Holothuroidea, and Crinoidea), followed by crus-taceans (Isopoda, Amphipoda, and Decapoda) and molluscs (Lamelli-branchia and Gastropoda). Less common hosts are annelids (Polychae-ta), arachnida (Xiphosura), and sipunculids, and a very small number of species form associations with coelenterates (Hydrozoa and Antho-zoa), other turbellarians (Alloecoela) and lower vertebrates (Elas-mobranchii and Teleostei). One obvious feature common to the major-ity of these symbioses is the occurrence of familial "host-type" specificity, in that the members of any one family tend to form asso-ciations with only one type of host organism. In the Rhabdocoela, for example, the family Umagillidae is exclusively symbiotic and all of its known members live in association with echinoderms (Echinoidea, Holothuroidea, and Crinoidea), except that one genus *Collastoma* lives in sipunculids. Even in families which are not wholly symbiotic the same tendency to associate with only one or two types of host can be found. Thus, in the Acoela, symbiotic members of the Convolutidae (*Aphanostoma* spp.) occur only in holothurians, and members of the Anaperidae (*Avagina* spp.), in echinoids. This association of a natural group, or taxon, of flatworms with a similar grouping of host species is, of course, seen also in some of the Trematoda and Cestoda. Hargis (1955, 1957), recognizing this phenomenon especially in mono-geneans, applies the term "supraspecificity" to these associations as opposed to "infraspecificity" where a single monogeneid species occurs on members of a single fish taxon). When the latter is a species then the infraspecificity shown is, of course, a species-specificity; if a genus, then it is a genus-specificity, and so on. These terms, as defined by Hargis, are most convenient and it is unfortunate that they have not been more widely adopted.

TABLE 1
Summary of turbellarian families in the orders Acoela, Alloeocoela,
Tricladida and Polycladida containing symbiotic genera*

Family	No.of known symbiotic species	Nature of symbiosis	Type of Host
ACOELA			
Anaperidae	3	entocommensal	Echinoidea
Convolutidae	2	"	Holothuroidea
Hallangiidae	1	"	"
Octocelididae	1	"	"
Nemertodermidae	1	"	"
Ectocotyla (?)	1	ectocommensal	Crustacea (Decapoda)
ALLOEOCOELA			
S.O.Archoophora	O		
S.O.Lecithoepithelia	O		
S.O.Cumulata			
Cylindrostomatidae	1	entocommensal	Lamellibranchia
Hypotrichinidae	4	facultative entocommensal;	Lamellibranchia;
		ectocommensal;	Crustacea (Leptostraca)
		parasitic	Teleostei
Plagiostomidae	1	ectocommensal	Crustacea (Isopoda)
S.O.Seriata			
Monocelidae	1	temporary shelter association	Gastropoda; Crustacea (Cirripedia)
TRICLADIDA			
S.O.Paludicola	O		
S.O.Terricola	O		
S.O.Maricola			
Bdellouridae	4	ectocommensal	Arthropoda (Xiphosura)
Procerodidae	1	"	"
Micropharyngidae	2	"	Elasmobranchii
POLYCLADIDA			
S.O.Acotylea			
Apidioplanidae	1	ectocommensal	Anthozoa
Emprosthopharyngidae	2	"	Crustacea (Decapoda)
Hoploplanidae	1	entocommensal	Gastropoda
Latocestidae	1	"	Lamellibranchia
Leptoplanidae	2	entocommensal(?)	Ophiuroidea; Amphineura
Stylochidae	2	ectocommensal;	Crustacea (Decapoda);
		entocommensal	Echinoidea
S.O.Cotylea			
Prosthiostomidae	1	ectocommensal	Crustacea (Decapoda)

*(from Jennings, 1971).

TABLE 2
Summary of turbellarian families in the order Rhabdocoela containing
symbiotic genera[*]

Family	No.of known symbiotic species	Nature of symbiosis	Type of Host
RHABDOCOELA			
S.O.Notandropora	0		
S.O.Opisthandropora	0		
S.O.Lecithophora			
Dalyellioida			
Acholadidae	1	parasitic	Asteroidea
Fecampiidae	6	"	Crustacea (Decapoda, Amphipoda, Isopoda); Annelida
Graffillidae	7	ectocommensal; entocommensal; parasitic(?)	Gastropoda; Lamellibranchia
Provorticidae	1	parasitic	Turbellaria
Pterastericolidae	2	entocommensal; parasitic	Asteroidea
Umagillidae	35	entocommensal	Echinoidea; Holothuroidea; Crinoidea; Sipunculida
Typhloplanoida			
Typhloplanoidae	1	ectocommensal	Polychaeta
Kalyptorhynchia	0		
S.O.Temnocephalida			
Actinodactylellidae	1	ectocommensal	Crustacea (Decapoda)
Craspedellidae	1	"	"
Scutariellidae	3	parasitic(?)	"
Temnocephalidae	31	ectocommensal	Crustacea (Decapoda); Gastropoda; Chelonia; Hydromedusae

[*](from Jennings, 1971).

Acoela

The Acoela are small flatworms, rarely longer than a few millimeters; they are entirely marine, lack an excretory system and a permanent gut lumen, and are generally regarded as the most primitive
living turbellarians. Only a few genera (Table 1) have species which
live symbiotically with higher organisms, but these come from five of
the six families generally recognized within the order. The only
ectocommensal is *Ectocotyle paguri* which Hyman (1951) describes as

living ectocommensally on hermit crabs. This species, the only one
of its genus so far described, has a simple caudal sucker which is
clearly an adaptation to its mode of life but nothing is known of its
relationships with its host. The precise taxonomic status of *E.
paguri* is doubtful and its acoel affinities, in fact, have been ques-
tioned. The reproductive system and plicate pharynx are more indicative of affinities with the Alloeocoela, and de Beauchamp (1961) places
it in this order.

Apart from *E. paguri* there are eight known symbiotic species of
acoels and these are all supraspecific with echinoderms, mainly holo-
thureans, and inhabit the gut or body cavity of their hosts (Jennings
1971). Unfortunately, as in the case of many other symbiotic Turbellaria, the published accounts of these species are restricted to
taxonomic descriptions and little is known of their general biology.
The indications are, though, that they are entocommensal rather than
parasitic and that they feed mainly on other organisms which share
their habitat within the common host (p. 137).

Rhabdocoela

The Rhabdocoela are small flatworms which live in marine, brackish, fresh water and occasionally damp terrestrial environments. They
fall into four suborders, and symbiotic species are found in the
Lecithophora (where they occur almost entirely in the marine Dalyel-
lioida except for one instance in the Typhloplanoida), and the Temnocephalida (Table 2). Eleven families in these two suborders of rhabdocoels have symbiotic members and although this is less than half of
the total number of turbellarian families which show this trait, the
number of species involved (at least eighty-nine) is significantly
larger than the total of symbiotic species (thirty-two) known from
the remainder of the Turbellaria (Jennings 1971).

The Lecithophora:Dalyellioida contains six families with symbiotic representatives; all of these are marine, live within their hosts,
and a number are truly parasitic. Perhaps the best known of these
are the Fecampiidae, parasitic in the body cavity of decapod, isopod,
and amphipod crustaceans and myzostomid annelids, and the Umagillidae,
commensal in the gut and body cavity of echinoderms and sipunculids.

Three genera occur in the Fecampiidae, i.e., *Fecampia, Kronborgia,* and *Glanduloderma.* *Kronborgia* is unusual in being one of the
few dioecious Turbellaria so far described. Among these it is unique
in being sexually dimorphic, with the males being only one-fourth to
one-sixth the size of the females. In all three genera the adult
worms lack a mouth and pharynx and in *Kronborgia* the intestine is
also absent.

The Umagillidae, with fourteen genera and thirty-five species,
all entocommensal, contain the majority of the symbiotic dalyellioid
rhabdocoels. The others occur, sparingly, in the Acholadidae (one
species), Graffillidae (three genera with seven species), Provorticidae (one species), and the Pterastericolidae (two genera each with
one species). Of these, the Graffillidae and Provorticidae also

contain free-living species.

The only instance of symbiosis in the Lecithophora:Typhloplanoida is *Typhlorhynchus nanus* (Typhloplanoidae), which is marine and lives on the body surface of the polychaete, *Nephthys scolopendroides* (Laidlaw 1902). The posterior end of the animal is flattened and expanded, and used for adhesion, but it is not organized into a definite sucker. The anterior end is produced into a type of tactile snout, which bears small papillae, and lacks the eyes and otolith which are generally present in related genera. Nothing is known of the general biology of *T. nanus* or of the degree of its dependence on the host.

The Temnocephalida are regarded here as a suborder of the Rhabdocoela, because of their very definite dalyellioid features and following Fyfe (1942) and Hyman (1951). They are all symbiotic and live in fresh water, in contrast to the symbiotic lecithophoran dalyellioids which are exclusively marine, and are the only turbellarians from this habitat to have formed symbiotic relationships with higher organisms.

The suborder contains four families and at least thirty-six known species (Table 2); the largest family is the Temnocephalidae with three genera and thirty-one species (Baer 1931; Jennings 1971; Schaefer 1971). Apart from the Fecampiidae, the temnocephalids are structurally the most modified of the symbiotic turbellarians, but, as will be seen later, physiologically they show few differences from typical free-living flatworms.

The temnocephalids live mainly on decapod or isopod crustaceans, and occasionally on other crustaceans, turtles, molluscs, and, rarely, on freshwater medusae. They occur on the gills, the lining of the branchial cavity, or the general body surface of the host. Geographically, they have a most interesting distribution. Found mainly in Australia, New Zealand, and Central and South America, species also occur in India, Ceylon, Madagascar, Indonesia, and some islands of the South Pacific. A few species, of somewhat atypical morphology, occur sparingly in the Balkans.

Baer (1951) sees this distribution as indicative of a very early origin for the temnocephalid-crustacean symbiosis. The parastacid crustaceans, which are the typical hosts in these associations, evolved in the early Cretaceous era during which the land masses of the Southern Hemisphere are believed to have been united by the Palaeoantarctic continent. These were subsequently separated by the oceans in the middle of the Tertiary period to give the modern continental pattern. Despite their geographical separation the temnocephalid-crustacean symbioses found in Australia and New Zealand are still remarkably similar to those of South and Central America and all the evidence points to a common origin before the land masses separated. There has been no independent evolution along divergent paths, and no indication of a move towards parasitism.

Alloeocoela

The Alloeocoela are small to medium-sized flatworms, rarely reaching 1 cm in length, and they are predominantly marine in habit. Some species, however, extend into brackish and fresh waters, and a few live in moist terrestrial habitats. Of the four suborders only one, the Cumulata, contains species which are truly symbiotic (Table 1), but instances of short-term associations in which the alloeocoel seeks temporary shelter within another organism occur in the suborder Seriata (p.132).

Three families of the Cumulata have symbiotic species, and these are all marine. In the Cylindrostomatidae one species, *Cylindrostoma cyprinae*, lives on the gills of various lamellibranchs in European waters (Hyman 1951), but its exact status is unknown. The Hypotrichinidae include four symbiotic species and these are of interest since they do not show the supraspecificity that is a characteristic feature of many turbellarian symbioses. The hosts include lamellibranchs, a leptostracan crustacean, and teleosts, and the symbioses themselves vary from apparently facultative commensalism to definite parasitism. *Urastoma frausseki* has been found on the gills of *Mytilus* and *Modiolus* (Dörler 1900) and also as a free-living species among algae and detritus (Westblad 1955). *Hypotrichina (Genostomum) tergestinum* and *H. marsiliensis*, in contrast, are ectocommensal on the leptostracan crustacean *Nebalia* in the Mediterranean (von Graff 1904-1908; Hyman 1951) and *Icthyophaga subcutanea* is one of the few parasitic turbellaria. It is also one of the very few which are symbiotic with vertebrates, and it lives subcutaneously in the branchial and anal regions of the teleosts *Bero* and *Hexagramma* (Syriamiatnikova 1949).

The remaining alloeocoel family with a symbiotic species is the Plagiostomaidae. *Plagiostoma oyense* lives on the general body surface of the isopod crustacean *Idotea*, and the symbiosis seems to be a permanent one since different sizes of individuals are found and the eggs are laid in cocoons which are cemented to the host's cuticle (de Beauchamp 1921; Naylor 1952, 1955).

Tricladida

The Tricladida are fairly large flatworms, generally 1 to 1.5 cm in length but in some species reaching up to 50 cm. They live in marine, brackish, freshwater, and terrestrial habitats, but only the marine suborder (Maricola) contains symbiotic representatives. These come from three families. The Bdellouridae, *Bdelloura candida, B. wheeleri, B. propinqua,* and *Syncoelidium pellucidum* inhabit the gill lamellae, branchial chamber, or general body surface of the horseshoe crab *Limulus* (Girard 1850; Wheeler 1894; Wilhelmi 1909). A single species of *Ectoplana*, from the predominantly free-living Procerodidae, is reported to have the same habitat (Kaburaki 1922). The other two species of symbiotic triclads come from the family Micropharyngidae and both are species of the single genus *Micropharynx*

M. parasitica and *M. murmanica* are found on the dorsal surfaces of the skates *Raja batis, R. clavata,* and *R. radiata* in the North Atlantic (Jägerskiöld 1896; Averinzev 1925), but nothing is known of their mode of life, nutrition, or life cycle.

Polycladida

The Polycladida are generally quite large flatworms. Most species reach several centimeters in length and are broadly oval in shape; they also show extreme dorsoventral flattening. They are all marine (except for a single species from fresh water) and are typically free-living predators. At least ten species, however, from seven families (Table 1) are reported to live symbiotically with other organisms, mainly hermit crabs and molluscs, and many others are commonly found sheltering in empty gastropod shells. *Emprostho-pharynx opisthopora* and *E. rasae* from the Emprosthopharyngidae, *Stylochus zebra* (Stylochidae) and *Euprosthiostomum* sp.(Prosthiostomidae) all live in association with hermit crabs (Bock 1925; Hyman 1951; Prudhoe 1968); and *Hoploplana inquilina* (Hoploplanidae), *Taenioplana teredini* (Latocestidae) and *Stylochoplana parasitica* (Leptoplanidae) live on the mantle or gills of various molluscs (Marcus 1952; Hyman 1951, 1967; Kato 1935a). Two other species are associated with echinoderms, *Discoplana (=Euplana) takewakii* (Leptoplanidae) lives in the genital bursae of ophiuroids (Kato 1935b) and *Discostylochus parcus* (Stylochidae) in echinoids (Bock 1925), while *Apidioplana mira* is reported to live symbiotically on the surface of the gorgonian *Melitodes* (Bock 1926).

Other polyclads live either on or in animals upon which they feed, but as the latter are eventually killed and totally consumed by the flatworms, these instances are best regarded as specialized types of predation. Some species of *Stylochus*, for example, creep between the shell valves of oysters and remain there for many days, gradually consuming the tissues and eventually killing the oyster (Pearse and Wharton 1938). Other species, like *Cycloporus papillo-sus*, live attached to flattened encrusting colonial tunicates such as *Botryllus* and *Botrylloides*, and feed by inserting the pharynx into the colony and sucking out individual zooids (Jennings 1957).

THE NATURE OF TURBELLARIAN SYMBIOSES

The various types of associations formed between the Turbellaria and higher organisms cover a very broad spectrum and are often difficult to define in strict terms, partly because so little is known about many of the individual symbioses. The symbioses have evolved, apparently independently, several times throughout the class but only a surprisingly small number have culminated in the evolution of a truly parasitic habit.

Shelter Associations

At one end of the spectrum of turbellarian symbioses, and grading imperceptibly into the life styles of the free-living predatory

species, are found associations which are not obligatory for either
partner since both can, and do, survive and reproduce in the absence
of the other. The important component of these associations seems
to be the "shelter" factor. Most of the free-living flatworms show
a predilection for sheltered, secluded situations and often creep
into crevices or among debris, into empty mollusc shells or beneath
stones, where they rest between the intervals of searching for food.
Occasionally, flatworms will seek shelter inside a shell still occupied
by the original owner. For example, the alloeocoel *Monocelis* (suborder Seriata) resists desiccation during the intertidal period by
creeping beneath the shell and into the pallial gills of the gastropod *Patella*, or between the shell valves of the barnacle *Balanus*.
When the "host" is submerged by the incoming tide, the flatworm
leaves its shelter and resumes its normal life. Similarly, acoels
and rhabdocoels normally regarded as free-living species are often
found among the byssus threads of *Mytilus* and *Modiolus*, or within
the mantle cavity of these and other lamellibranch or gastropod molluscs. The reported association of the alloeocoel *Urastoma frausseki*
with *Mytilus* and *Modiolus* may in fact be of this type, since it has
been found both as free-living individuals and as an apparent symbiont in the mantle cavity of its supposed hosts.

In all the associations with lamellibranchs, and also in those
involving the filter-feeding gastropods, the shelter factor will
obviously be reinforced by a nutritional one. The feeding current
of the mollusc brings a supply of suitably sized food particles
or organisms into the mantle cavity and these could be utilized by
any flatworms living there. This can only be suggested for the various rhabdocoels and alloeocoels living in the mantle cavities of
filter-feeding molluscs until more is known of their diet. However,
it has been shown to be the case in the related rhynchocoelan *Malacobdella grossa* which lives commensally in the mantle cavity of
Zirfaea crispata and other lamellibranchs (Gibson and Jennings 1969).
M. grossa, however, shows considerable modification of the feeding
mechanism away from the type characteristic of the free-living predatory rhynchocoelans, and has evolved its own type of filter feeding. Investigation of the feeding mechanisms of those turbellarians
living in molluscan mantle cavities would clearly be worthwhile, but
the taxonomic accounts available give no indication of any modifications of the buccal cavity or pharynx comparable to those found in
M. grossa.

Commensal Symbioses

The majority of turbellarian symbioses come under the general
subheading of commensalism, provided that this convenient term is
used merely to indicate an association in which neither of the partners profits at the expense of the other. Commensalism has become
increasingly used in this sense (Read 1970), rather than in its
older and philologically correct sense of implying the sharing of
the same food. As will be seen, the latter is indeed the case in a

number of instances but it is not universal. The Turbellaria are predominantly predators and this basic habit persists to varying extents in many of the symbiotic species, so that, although they may live in the host's gut or body cavity, they still feed on the same type of organisms, and in the same manner, as their free-living relatives.

Ectocommensal Symbioses

The ectocommensal habit has evolved a number of times within four of the five turbellarian orders, the only exception being the Acoela. The only acoel reported with this habit is *Ectocotyla paguri* and, as discussed earlier, there is some doubt on morphological grounds as to the validity of placing this species within the Acoela. Except for *E. paguri*, the acoel grade of organization does not appear to have been adaptable to ectocommensalism, but no obvious reasons for this can be suggested.

In the Rhabdocoela members of the suborder Temnocephalida are almost entirely ectocommensal. The only noncommensals appear to be the aberrant members of the Scutariellidae, which Mrazeck (1906) and Matjasic (1957) claim are parasitic. The temnocephalids live most commonly on the gills, the inner surfaces of the branchial chamber, and the general body surface of freshwater decapod crustaceans. In these situations, and especially in the branchial regions, the flatworms are in an excellent position to capture fragments of the host's food released from an original food mass by the shredding action of the crustacean's feeding mechanism and accidentally swept away from the mouth by the respiratory or other currents. It is plausible to suggest that this might be the basis of the association between temnocephalids and crustaceans, but in the few species that have been studied in detail, particles of the host's food seem to form only an insignificant proportion of the diet. Examination of gut contents indicate that temnocephalids such as *Temnocephala novaezealandiae*, *T. bresslaui*, and *T. brenesi* feed mainly on diatoms, protozoa, rotifers, nematodes, and small oligochaetes and crustaceans (Fyfe 1942; Gonzales 1949; Jennings 1968b), and have, therefore, precisely the same diet as the free-living rhabdocoels (Jennings 1968a). Even so, living in or near the respiratory current will bring a constant supply of food organisms to the turbellarians, even when the host itself is not feeding.

Thus, it is debatable whether the temnocephalids are commensal in the literal interpretation of the term, but they obviously derive some advantages from living in these particular locations on this type of host. Their retention of a feeding pattern typical of the free-living turbellarian predators, without dependence on the host's feeding activities, accounts for their ability to live in areas away from the respiratory current, such as the carapace and telson of crustacean hosts or the general body surface of turtles, gastropods, or, very rarely, hydromedusae.

Many of the details of temnocephalid symbioses have not been studied and these associations pose several intriguing problems.

138

First, it is by no means certain that the relationships are obligatory for the flatworm, although no free-living forms have yet been discovered. Some species have been maintained for several months away from their hosts and they feed and produce viable eggs quite readily (Gonzales 1949; Hickman 1967; Jennings 1968b). Second, there is the question of how those species which live on crustacean hosts and which constitute the majority of temnocephalids survive host ecdysis and manage to reestablish themselves on the newly molted host. They are generally very active and agile animals, capable of rapid locomotion in a leech like manner by looping the body and attaching to the substratum alternately by the tentacles and sucker. They may well move quickly off the cast exoskeleton back on to the host or even achieve the transfer during actual exuviation of the old cuticle. Alternatively, a brief or even extended free-living period may be a normal feature of the life cycle. Support for this possibility comes from the observation by Jennings (1968c) on egg laying in *T. brenesi*, commensal on the freshwater shrimp *Macrobranchium americanum* in Central America (Table 3). Fifty-five percent of the hosts examined

TABLE 3
Summary of data on the egg-laying habits of *Temnocephala brenesi*[*]

% hosts infested	Average no.of adults/host	Average no. of embryonated eggs	Development time of eggs in laboratory	Rate of egg-laying
55	8	150	21–24 days	1 in 3-4 days

[*](based on Jennings, 1968c).

were infested, with an average of eight temnocephalids per host. The gills usually bore clusters of egg capsules and many more eggs than adults were always found; in some cases, gills yielding only 3 adults carried more than 150 embryonated eggs. In the laboratory, development from the newly laid egg to emergence of the miniature adult occupies twenty-one to twenty-four days and sexually mature adults in vitro produce one egg every three to four days. Considering this rate of egg production and the time span of development against the numbers of adults normally found on any one host (examined immediately on collection), it is apparent that a larger number of adults per host could reasonably be expected than was ever found. Further, eggs containing very early embryos were sometimes found on gills which did not yield adults. Thus, it seems possible that at least this temnocephalid normally spends some time away from the host, or may even visit the host only to deposit its eggs. Unfortunately, free-living specimens were never found, despite intensive searches for these in the host's habitat.

The temnocephalid's habit of cementing its eggs to the host cuticle means that a proportion of them must sometimes hatch after the cuticle has been molted and, under these conditions, the fate of the newly emerged individuals is not known. Most species produce large numbers of eggs, however, so that even if such juveniles fail to reach a new host there will be no significant effect on the survival of the species.

In at least one species, this habit of laying eggs on the host creates conditions favoring the growth of food organisms which are utilized by the newly hatched juvenile. *T. brenesi* lays its egg capsules in rows of five to ten along the distal margins of the gill lamellae of its host. As soon as they are laid, small particles of detritus, carried over the gills by the respiratory current, gather between and around the capsules and form a substratum which in turn rapidly develops a rich growth of diatoms, protozoa, and rotifers. As soon as the young temnocephalid emerges from its capsule, it starts to feed on this material and rapidly grows to the adult size when it then assumes the adult feeding habits. It would be interesting to compare growth rates of the young *T. brenesi* with those of free-living rhabdocoels, whose young have to seek out their food organisms or rely on chance encounter during random wanderings. If the commensal reached maturity significantly earlier, then, this would demonstrate an inherent advantage in the mode of life and indicate a possible stimulus for its evolution.

Other than the Temnocephalida only one other rhabdocoel with the ectocommensal habit is known. *Typhlorhynchus nanus* (Lecithophora: Typhloplanoida, Table 1) lives attached to the body surface of the polychaete *Nephthys scolopendroides*, but nothing is known of its physiology or life history.

Ectocommensalism has evolved in some of the Alloeocoela and, in the few instances where the associations have been studied, indications are that they are basically the same as the temnocephalid symbioses. *Plagiostoma oyense*, for example, living on the surface of the isopod *Idotea*, cements its cocoons to the host's cuticle, and de Beauchamp (1921) reports that the intestine often contains empty rotifer cuticles. Thus, like the temnocephalids, *P. oyense* appears to be using its host principally as a substratum for egg deposition and a platform from which it feeds as a predator in the typical turbellarian manner. Other symbioses involving alloeocoels (*Hypotrichina tergestinum* and *H. marsiliensis* on the surface of *Nebalia*, and *Cylindrostoma cyprinae* on the gills of various lamellibranchs) are probably ectocommensal in nature, judging from the location of the flatworms on their hosts, but nothing is known of diets or breeding habits.

Clearly defined ectocommensalism, however, is found in three genera of the Tricladida, and in these there are several marked similarities with the temnocephalid symbioses. *Bdelloura, Syncoelidium*, and *Ectoplana* all live on the gill lamellae of the horse shoe crab, *Limulus*. The host has a similar feeding mechanism to that of the crustacean hosts of temnocephalids; shredded particles of its food accidentally drift back over the gills and are available to the triclads. Unfortunately the efficient disintegrating action of the triclad's feeding mechanism based on the plicate pharynx, which is virtually the same in these symbiotic species as in the free-living triclads, makes it almost impossible to identify the flatworm's intestinal contents and it has not been possible to determine the relative proportions in the diet of materials originating from the host's food supply. The number of triclads present on any one host is generally quite large, however, and considerably in excess of the number of temnocephalids found per host. Over 300 mature *Bdelloura candida*, for example, were found by the author on one large female *Limulus polyphemus* collected on the Massachusetts coast in August 1971, and it seems unlikely that adequate numbers of the organisms upon which triclads normally prey (small annelids, crustaceans, etc.) would be present in the respiratory current of the host to support this large number of symbionts. It is more reasonable to suggest that the host's food is the major source of nourishment for the triclads, but in the absence of the necessary data this possible interesting difference between the triclad and temnocephalid symbioses remains purely speculative.

As do the temnocephalids and the alloeocoel *Plagiostoma oyense*, all commensal triclads deposit their eggs in capsules which are cemented to the host's gill lamellae. According to Wheeler (1894), the different species select different sites on the same host for oviposition, and also vary somewhat in regard to breeding season. *Bdelloura candida* deposits its egg capsules fairly randomly over the entire surface of the gill lamellae, but *B. propinqua* selects the basal region and *Syncoelidium pellucidum* a narrow area near the edges of the lamellae. *B. candida* lays eggs in May and early June, when the host *Limulus* returns from deep water onto sandy beaches to breed, but *B. propinqua* and *S. pellucidum* lay eggs in late July or early August. It has been suggested that migration of the triclads between hosts occurs during breeding (Wheeler 1894), but since the host's eggs are left buried in the sand during development, it is not clear how young hosts become infested. This question, as well as how the symbionts survive host ecdysis, the precise nature of the symbiont's diet, and the general biology and ecology of the symbiosis with up to three species of symbionts living on the one host species offer an intriguing field for future investigation.

The two other symbiotic triclads, *Micropharynx parasitica* and *M. murmanica*, are mainly of interest in that, along with the alloeocoel *Icthyophaga*, they are the only turbellarians to have formed symbioses with vertebrates. No details of their life on their skate hosts are known, however, and while their reported habit of remaining attached to the dorsal surface suggests ectocommensalism, their precise status as symbionts remains undefined.

A number of the symbioses involving polyclads are probably of the ectocommensal type, but again there is little information available to confirm this. The polyclad symbioses probably stem from the polyclads' habit of seeking shelter in empty gastropod shells and they have developed along divergent lines. In one type of symbiosis, as with some of the rhabdocoels and alloeocoels like *Urastoma*, the polyclad utilizes shells still occupied by the rightful owner. In these organisms, the symbiosis is supraspecific rather than infraspecific. *Hoploplana inquilina*, for example, occurs in the mantle cavities of various gastropods including *Busycon canaliculatum*, *Thais haemastoma*, and *Urosalpinx cinerea* (Marcus 1952; Hyman 1967). In other symbioses the polyclads have become associated with hermit crabs which, like themselves have sought shelter in empty gastropod shells. In these symbioses the shelter factor for the polyclad would probably be rapidly reinforced by a nutritional one of the same type as suggested in the triclad-*Limulus* symbioses and, more tentatively, in the temnocephalid-crustacean ones. The hermit crab's feeding mechanism, which shreds the food into small particles, and the respiratory mechanism, which sets up currents in the gastropod shell, together would provide a polyclad sheltering in the shell with a supply of easily ingested food. The free-living polyclads are predators, of course, and not particulate feeders and so here again it would be interesting to know more about the feeding mechanisms of polyclads such as those species of *Stylochus*, *Emprosthopharynx*, and *Euprosthiostomum* which live symbiotically with various species of hermit crabs.

Some symbioses of polyclads and hermit crabs show a remarkable degree of infraspecificity, down to the level of species specificity, in contrast to the supraspecificity shown by polyclad-gastropod symbioses. *Emprosthopharynx rasae*, for example, lives in shells of *Trochus sandwichensis* occupied by the hermit crab *Calcinus latens*, but is never found in *T. sandwichensis* shells occupied by *C. laevimanus* or *Clibanarius zebra* (Prudhoe 1968). Here, some unknown factor must be responsible for the species specificity; quite simply, it could be that because *C. latens* extends further down the shore into the subtidal zone than do the other two species, the pelagic larvae of the polyclad preferentially establish themselves in shells occupied by this species.

Entocommensal Symbioses

The entocommensal Turbellaria are mostly either acoels or rhabdocoels and the great majority are symbiotic with echinoderms (Tables 1 and 2). One of the most significant features of these symbioses is that the diets of the symbionts, whenever they are known, are virtually the same as those of related free-living species. The free-living acoels and rhabdocoels feed on bacteria, diatoms, protozoa, and tiny annelids and crustaceans (Jennings 1968a, 1973). These organisms themselves, with the exception of annelids, commonly live as commensals inside echinoderms and especially in echinoids and holothurians (Hyman 1955). They occur in these hosts, in fact, to a much greater extent than in any other invertebrate types, and it is clearly this ready availability of food that has led to the dominance of echinoderms as hosts for entocommensal turbellarians. The flatworms have been able to colonize a new type of habitat, within another organism, without changing the diet or feeding mechanism. Thus, acoels which are entocommensal in echinoderms retain a pattern of nutrition similar to, if not identical with, that of free-living species. Westblad (1949), studying *Meara stichopi* from the intestine and body cavity of the holothurian *Stichopus tremulus*, observed the

remains of copepods and diatoms in the acoel's gut and deduced from
the appearance of preserved specimens that this species feeds by pro-
truding a portion of the intestine through the mouth and engulfing
its food in an amoeboid fashion. The free-living *Convoluta convoluta*
feeds in precisely this way (Jennings 1957). Similar gut contents
have been reported for other entocommensal acoels as illustrated by
Hickman (1956) who described diatoms in the gut of *Avagina vivipara*
living in the esophagus of *Echinocardium cordatum*.

In the entocommensal rhabdocoels the same pattern is found. In the
Umagillidae *Syndesmis antillarum* from the gut and coelom of *Lytechi-
nus variegatus*, and *S. franciscana* from the same sites in *Stronglyo-
centrotus franciscanus* and *S. purpuratus*, feed on the ciliate protozoa which live symbiotically in these hosts (Jennings and Mettrick
1968; Mettrick and Jennings 1969). *S. antillarum* on occasion also
ingests host coelomocytes, but this is apparently by chance and the
coelomocytes do not form a significant proportion of the diet. In
both species the digestive physiology, as the feeding habits, resem
bles that of the free-living flatworms and the only major difference
in the general pattern of nutritional physiology is that the symbiotic species show considerable emphasis on glycogen storage (p. 144).

In the Graffillidae there are two symbiotic genera, *Graffilla*
and *Paravortex*. Their exact status has not been defined, but it
seems likely that the former is parasitic (p. 144), while the two
known species of *Paravortex* seem to be midway between being entocommensal and being parasitic. *P. cardii*, living in the stomach of
Cardium edule, generally has diatoms among its gut contents and sometimes considerable quantities of mucus (Jennings, unpublished observations). The diatoms are of the same type as those present in the
host's stomach, so the flatworm is utilizing the host's food and
feeding mechanism, and at least a portion of the mucus is probably
of host origin and derived from that used to collect the food. Often,
the free-living flatworms secrete mucus to facilitate the capture
and ingestion of the food; thus, some mucus seen in the intestine of
P. cardii may be endogenous in origin. *P. gemillipara* occurs on the
gills and in the kidney of *Modiolus demissus (=M. plicatulus)* and
specimens from the gills often have diatoms and mucus in their intestine (Jennings, MacLeod, and Osman, unpublished observations). Specimens from the kidney, however, did not contain anything recognizable
and their diet remains unknown. During examination of the host for
P. gemillipara, two unknown species of *Macrostomum* were often found
on the gills or in the mantle cavity. These three species were also
commonly found among the host's byssus threads and were free swimming
in the general habitat, illustrating the point made earlier as to the
occurrence of normally free-living rhabdocoels and acoels in habitats
adopted by other species that have become entirely entocommensal.

Parasitic Symbioses

Remarkably few symbiotic Turbellaria are parasitic in the sense
of living partially or entirely at their host's expense. In some,

as in the rhabdocoel family Fecampiidae, the symbioses have received considerable attention and their parasitic nature is obvious, but in others very little is known of the general biology of the symbiont and its parasitic mode of life has been deduced from structural features such as the reduced size, or absence, of the mouth, pharynx, or intestine.

The effect which the parasite's presence has on the host varies in the different symbioses. The extreme situation is found in *Kronborgia*, in the Fecampiidae. *K. amphipodicola* is parasitic in the tube-building marine amphipod *Amphiscela macrocephala* (Christensen and Kanneworff 1965) and the males and females of this dioecious species pass the greater part of their life in the host's body cavity. They lack eyes, mouth, pharynx, and intestine, and presumably absorb their nourishment through their body wall from the host's body fluids. They eventually emerge from the host at the posterior end. The female secretes a cocoon around herself which the male enters to fertilize her, then leaves it again to die soon afterwards. The female eventually deposits several thousand eggs in the cocoon before she too vacates it and dies. Each egg gives rise to a ciliated larva which seeks out a new host amphipod, encysts on the cuticle, and eventually bores through this and enters the haemocoel to complete the cycle. During growth of the rhabdocoels to maturity the host's gonads atrophy, presumably due to successful competition by the flatworms for the available nutrients, and both sexes of the host are rendered sterile as a consequence. Then, as the mature *Kronborgia* start to leave the host by boring through the body wall, the amphipod suddenly becomes immobile and dies. Emergence of the female flatworm, particularly, may take several minutes and Christensen and Kanneworff (1965) interpret this initial paralysis and then death of the host as being advantageous to the parasite as it prevents damage which might result from the host's movements. This may well be so, but both it and the preceding parasitic castration of the host would appear to reflect an incomplete adaptation in the symbiosis. Further, it presupposes that in multiple infections of any one host, all the parasites must be ready to emerge at about the same time; otherwise, any immature ones not yet ready for emergence, cocoon secretion, and mating must surely die soon after the host.

Other members of the Fecampiidae similarly live in the haemocoels of crustacean hosts, but the partners in these symbioses, unlike *Kronborgia* and its host, seem to be well adapted for life together and no adverse effects have been reported. *Fecampia erythrocephala* occurs in the decapods *Cancer pagurus*, *Pagurus bernhardus*, and *Carcinus maenas* (Giard 1886), *F. xanthocephala* in the isopod *Idotea neglecta* (Caullery and Mesnil 1903), and *F. spiralis* is found in the isopod *Serotis schytei* Baylis 1949). The life cycles are similar to that of *Kronborgia* in that cocoons are laid after the flatworm vacates the host within which it has reached maturity, but numerous cocoons are laid instead of a single one. Each cocoon, however, is much smaller than the adult flatworm and contains only two eggs. A ciliated larva develops from each egg, seeks out a new host, and penetrates into the haemocoel. Further points of difference from *Kronborgia* are that these species of *Fecampia* are hermaphrodite and their motile larval stages show a closer resemblance to the free-living rhabdocoels in that they possess eyes, mouth, pharynx, and intestine. Except for the intestine, these structures are lost, however, when entry is gained into a new host. Since *Kronborgia* lacks these structures at all stages of the life cycle, it might be thought that it is more highly adapted to the parasitic life style, or that its association with its host is more ancient, but its habit of castrating the host and eventually killing it would not appear to be concomitant with either of these interpretations. The third and remaining genus in the Fecampiidae, *Glanduloderma*, is also described as parasitic (Jagersten 1942). *G. myzostomatis* lives in the mesenchyme of the myzostomid annelids *Myzostomum brevilobatum* and *M. longimanum*; it lacks mouth and pharynx but details of its life cycle are unknown.

Only a few of the other symbioses in which turbellarians live parasitically show a deleterious effect upon the host which is comparable in some measure to that caused by *Kronborgia*. They never directly cause the death of the host, but parasitic castration or reduced growth rates have been observed. In the rhabdocoel family Provorticidae, for example, *Oikiocolax plagiostomorum* is the only parasitic representative and lives in the mesenchyme of the marine alloeocoel *Plagiostomum* (Reisinger 1930). Infected hosts consistently show degeneration of the ovaries but Reisinger does not record the symbiosis as ever causing the death of the host. In the Pterastericolidae a recently described species, *Triloborhynchus astropectinis*, from the pyloric caeca of the starfish *Astropecten irregularis* feeds on the caecal contents and occasionally the caecal tissue and there is some evidence that its presence retards the growth of the host (Bashiruddin and Karling 1970).

The other parasitic turbellarians appear to have much less effect on their hosts but this may simply be an indication of our lack of knowledge about these particular symbioses. These include a single species in the Acholadidae, *Acholades asteris*, which lives encysted in the connective tissue of the tube-feet of the starfish *Coscinasterias calamaria* and lacks eyes, mouth, pharynx, and intestine (Hickman and Olsen 1955); and at least five species from the Graffillidae which live mainly in the kidney and kidney ducts of marine gastropods (von Jhering 1880; von Graff 1904-1908; Dakin 1912). *Graffilla buccinicola*, from the whelk *Buccinum*, is of especial interest here as Dakin reports it from the mantle cavity, apparently living as a commensal, as well as from the kidney. Infected hosts from the Massachusetts coast may contain very large numbers of *G. buccinicola* distributed virtually throughout the body, but without suffering any apparent ill effects.[*]

The temnocephalid rhabdocoels are almost entirely commensal and have already been considered under that heading. One aberrant family

---

[*]M. P. Morse: personal communication.

of the Temnocephalida, however, is the Scutariellidae. Species of
Scutariella, from the Balkans, possess the typical temnocephalid
habit of living on the gills of freshwater decapod crustaceans, but
they are reported to feed on gill tissue supplemented by the host's
blood which they suck in by means of the doliiform pharynx (Mrazeck
1906; Matjastic 1957). The Scutariellidae include two other genera,
Monodiscus and *Caridinicola*, from Ceylon and India, and it would be
interesting to discover their feeding habits.

The remaining known parasitic turbellarian is the alloeocoel
Icthyophaga subcutanea. This species lives subcutaneously in the
anal and branchial regions of the teleosts *Bero* and *Hexagramma*, and
is described as being generally red in color (Syriamiatnikova 1949).
The gut occupies most of the body and thus the red coloration probably comes from ingested host blood, as is often the case with trematodes. The pharynx is of the doliiform type, highly muscular and
suitable for rupturing small vessels and sucking in blood. Although
the parasite lives subcutaneously, the epidermis is uniformly ciliated and eyes are present anteriorly, suggesting that the symbiosis
or at least the subcutaneous burrowing may be of relatively recent
origin. In this connection it is perhaps significant that *Icthyophaga* is found only in the branchial and anal regions, indicating
perhaps that the branchial chamber and cloaca were the original habitats, and that the subcutaneous habit has evolved from living in
these protected cavities. The idea that either the entire symbiosis
or the subcutaneous habit, is of recent origin is supported by the
fact that the flatworm is often found within cysts of host fibrous
tissue, indicating some host reaction to the presence of the parasite.

STRUCTURAL, PHYSIOLOGICAL, AND LIFE CYCLE
MODIFICATIONS IN SYMBIOTIC TURBELLARIA

Structural Modifications

It might be expected that major structural modifications would
be found in the body surface of the symbiotic turbellarians to fit
it for its role as the host-symbiont interface, but there are remarkably few instances of this. Even the parasitic species retain an
epidermis very similar in appearance to that of the free-living
species.

The turbellarian epidermis is typically a single-layered structure, cellular or syncitial, often pigmented, and sometimes with the
nuclei "insunk" into the underlying mesenchyme. It is generally
ciliated, with short microvilli visible at the ultrastructural level
between the cilia, and contains scattered mucus gland cells. The
epidermis is perforated by the necks of various subepidermal gland
cells whose secretions are discharged onto the body surface. Its
cells are generally loaded with rhabdites, rod-shaped bodies which

originate in mesenchymal cells and pass through the epidermis onto the body surface. Here they hydrate and fuse into a gelatinous "fluid cuticle" which is protective yet compatible with ciliation (Jennings 1957).

The most profound departure from this typical turbellarian epidermis is found not in the entocommensal or parasitic forms as might be expected, but instead in the ectocommensal temnocephalids. In these forms the epidermis is syncytial and often has insunk nuclei, as in some free-living species. The distal border, however, has become modified into a clear, tough, cuticle like structure which is present over the entire body, and as a consequence cilia are either very sparse or, more usually, absent. Pigmentation is reduced or absent, and rhabdites are few in number and present only anteriorly where they are associated with the tentacles. Mucus glands, in contrast, are concentrated posteriorly and supply the adhesive organ.

The functional significance of these modifications of the temnocephalid epidermis is unknown. They have apparently not been necessary in those triclads which live in a strictly comparable habitat on the gills of *Limulus*, nor in entocommensals such as *Paravortex* which live in the alimentary canal of their hosts. The temnocephalids are the only symbiotic turbellarians from fresh water; thus, some osmotic factor may be involved. However, it is difficult to see why this factor would not also operate on the free-living freshwater flatworms which have a normal turbellarian type epidermis.

Modifications of the epidermis in other commensal turbellarians, when they occur, usually take the form of reduced ciliation and pigmentation as well as reduction in the numbers of rhabdites and epidermal mucus glands. Occasionally there is an apparent thickening of the epidermis due to an increase in the height of the cells, but this and the other features listed also occur in some free-living species. The entocommensal acoel *Meara*, for example, shows all these features, but so too does its free-living close relative *Nemertoderma* (Westblad 1937).

The parasitic species show the same slight tendencies toward the loss of pigment, rhabdites, and cilia as do the commensal ones, but again there are few extreme conditions. *Icthyophaga*, although living subcutaneously on its host, retains a normal ciliated epidermis. In contrast, the rhabdites and mucus glands have been lost in the rhabdocoel *Acholades*. In *Triloborhynchus* these are present, but ciliation is reduced and confined to the anterior end. *Pterastericola*, however, in the same family as *Triloborhynchus* and living in the same habitat within starfishes, retains an almost typical turbellarian epidermis (Karling 1970). The Fecampiidae retain cilia and rhabdites, but in the one species studied in detail, *Kronborgia amphipodicola*, the microvilli which occur between the cilia are both longer and much more numerous than those in free-living flatworms (Bresciani and Køie 1970). These authors also describe the discharge from the epidermis of vesicles of secretion which might be enzymic in function, or alternatively, concerned with cocoon formation. In the absence of a gut, *Kronborgia* presumably absorbs its nutrients across the epidermis and the microvilli, by providing an enormous increase in surface area must greatly facilitate this process. It is possible, too, that if the secretions from the epidermis are enzymic, the microvilli may also be concerned with membrane or "contact" digestion, of the type first described in the mammal by Ugolev (1960, 1965) and subsequently suggested for a number of other animal groups including trematodes and cestodes (Halton 1966; Jennings 1968a, 1972; Smyth 1969).

Adhesive organs show a certain amount of modification and elaboration in the symbiotic turbellarians, but again not to the extent that might be expected. The majority of the symbionts continue to utilize the basic turbellarian method of adhesion by means of mucous secretions, often discharged through or onto special adhesive papillae. (The question of how adhesion is terminated, when the flatworm wishes to release its hold on the substratum, has not been satisfactorily answered.) Mucuous adhesion is supplemented, or replaced, in some symbiotic genera by the development of simple glandulomuscular folds, as in members of the Pterastericolidae and in *Typhlorhynchus* (Typhloplanidae), or by the organization of these into definite, cup-shaped, muscular suckers. Such suckers, which are not as elaborate as those of trematodes and cestodes, occur in the aberrant acoel *Ectocotyla*, the temnocephalids, the ectocommensal triclad *Bdelloura* and in the cotylean polyclads such as *Cycloporus*. As discussed earlier, cotyleans like *Cycloporus* are not symbiotic in any strict sense, and the occurrence in them of a simple muscular sucker is one more example of how virtually all features of the symbiotic Turbellaria can be found elsewhere in free-living predatory species.

Other than the extreme instances of the loss of part or the whole of the alimentary system in the adult stages of the parasitic rhabdocoels *Acholades, Oikiocolax, Fecampia, Kronborgia,* and *Glanduloderma,* the symbiotic Turbellaria show surprisingly few internal structural modifications that can be correlated with their mode of life. All of these symbionts presumably absorb their nourishment across the body wall, and in this respect resemble the cestodes. In contrast, the only other parasitic turbellarian, *Icthyophaga*, retains a well-developed gut with a muscular bulbous pharynx and appears to be sanguivorous like many trematodes. Similarly, the most externally modified of the symbiotic turbellarians, the ectocommensal temnocephalids, with their unusual epidermis and posterior sucker, show virtually no internal modifications and the gut in particular is very like that of free-living species.

The female portion of the reproductive system is the only other internal structure showing any modifications which can be related to the symbiotic habit. No new structures have evolved, but there is sometimes an emphasis on increased egg production which is shown by elongation of the ovaries (relative to those of free-living species), and elongation and branching of the yolk glands. Such modifications are seen in the rhabdocoel family Umagillidae. In *Syndesmis*, for example, the ovaries and yolk glands of a mature specimen can occupy up to two-thirds of the body. The umagillids and temnocephalids produce more eggs, and at a greater rate, than do free-living species, and this has been reported also for the symbiotic acoel *Avagina glan-dulifera* (Westblad 1953).

Physiological Adaptations

The physiology of the symbiotic Turbellaria has not been extensively investigated. Studies have been confined to the temnocephalid and umagillid rhabdocoels, and these have mainly been concerned with nutritional physiology and food reserves.

Where the symbionts retain a functional alimentary system, indications are that the diet, gut structure, and digestive physiology are virtually the same as in free-living turbellarians. Thus, while the ectocommensal temnocephalids *Temnocephala brenesi* and *T. novae-zealandae*, and the entocommensal umagillids *Syndesmis antillarum* and *S. franciscana* show some minor variations in the structure of their gastrodermis, as compared with the free-living triclads or rhabdo-coels, their diet, the site and sequence of their digestive processes, and the types of digestive enzymes involved are very similar to those of the nonsymbiotic turbellarians (Jennings 1968b; Jennings and Mettrick 1968; Mettrick and Jennings 1969; Jennings 1973).

In those parasitic symbionts in which the gut is either absent or manifestly nonfunctional, the pattern of nutrition must vary significantly from that which is characteristic of the commensal and free-living species and approach that seen in the cestodes. Unfortunately, the nutritional physiology of these genera has not been studied; an investigation of *Fecampia* or *Kronborgia* along the lines used to study cestode nutrition should yield interesting results and it is hoped that such a study will eventually be made. The increase in the number and size of the epidermal microvilli in *Kronborgia* has already been noted and the role of the epidermis in nutrition would be a useful starting point.

Although the entocommensal turbellaria show no significant modifications in their digestive physiology, there are interesting differences in their food reserves as compared with those of the free-living species, and especially with regard to the amount of glycogen stored. A comparison of the percentage of glycogen in the dry weight of the umagillid *Syndesmis*, for example, with that of other platyhelminths shows a definite shift in *Syndesmis* away from the typical free-living turbellarian pattern and toward that of entoparasitic trematodes and cestodes (Fig. 1). Four free-living species had an average of 9.96% glycogen in the dry weight (S.E. ± 2.01), while the values for two species of *Syndesmis* were 17.73% (± 1.8); for nine species of digenetic trematodes, 14.77% (± 2.88); and for twenty species of cestodes, 26.99% (± 3.0). Although the difference between the values for the trematodes and the free-living turbellarians is perhaps not statistically significant since the two standard errors show a slight overlap (Fig. 1), nevertheless the values do show a definite trend toward increased glycogen storage with assumption of the symbiotic life style. The original trematode data included a very low value (3.5% glycogen/dry weight) for female *Schistosoma*

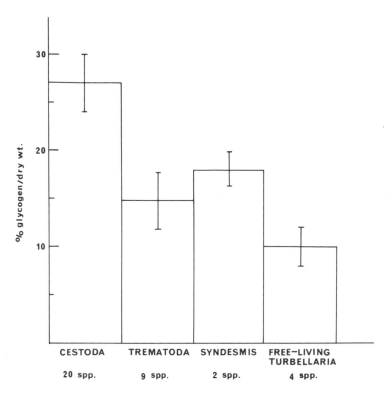

Figure 1. Histogram showing the mean % glycogen/dry weight for cestodes, digenetic trematodes, the entocommensal turbellarian
Syndesmis, and free-living turbellarians based on data from
Smyth 1966, 1969; Jennings and Mettrick 1968; Mettrick and
Jennings 1969; von Brand 1936; and Mettrick and Boddington
1972a. The original data, given as % glycogen/dry weight
for individual species by original authors, were trans-
formed to arcsines ($\Theta = \sin^{-1}[\,p\,]$) and retransformed to
the percentages shown in the histogram by $p = [\sin \Theta^2]$.
Confidence limits are two standard errors. This method
follows Sokal and Rohlf 1969.

mansoni. If this value is excluded, the mean trematode value rises
to 16.41% (± 2.70) and the trematode and turbellarian standard errors
no longer overlap. This use of these data may be challenged, of
course, on the grounds of the small number of samples of entocommensal and free-living turbellarians, but no further data are available.
    The high glycogen content of entoparasitic helminths has been
well known for some time, and attempts have been made to relate it
to the oxygen tension of the various habitats (von Brand 1966). The
inference has been that the high glycogen content indicates an emphasis on carbohydrate metabolism and the release of energy by anaerobic
glycolysis. This may well be correct in many instances, but in others
the entoparasites live in aerobic habitats, such as amphibian lungs
(Halton 1967). Similarly, the habitat of *Syndesmis* (the gut and
coelom of echinoids) is far from anaerobic, so that in these cases
some other factor must be operating. One feature common to the

entocommensal and the parasitic flatworms is that they both live within their hosts and therefore have the problems of the dispersal of their young stages and the infection of new hosts. They all show increased egg production as one answer to these problems, so there may well be a link between this increased fecundity and glycogen storage. The ectocommensal temnocephalids also show increased egg production, as compared with free-living species, but unfortunately no quantitative data on their glycogen content are available. Fernando (1945), however, reports that histological methods reveal concentrations of glycogen around the ovaries in two species from Ceylon and India, and Jennings (1968b) describes similar concentrations around the testes of a Central American species.

Only one other aspect of the physiology of symbiotic turbellaria has been investigated, namely, the molar ratios and concentrations of the free amino acid pools of the umagillid *Syndesmis franciscana* (Mettrick and Boddington 1972b). The amino acid pool of the umagillid differs in both array and concentration from the host coelomic fluid; it is similar in array but differs in concentration from those of other marine invertebrates (including turbellarians) and it differs in both array and concentration from those of freshwater turbellarians. Significantly, it is similar in both array and concentration to the amino acid pools of many cestodes (e.g., *Hymenolepis microstoma, H. diminuta,* and *Lacistorhynchus tenuis*). The metabolic significance of these similarities is as yet unknown, but they constitute a further resemblance between an entocommensal turbellarian and the wholly parasitic members of the same phylum.

Modifications in the Life Cycle

Life cycles in the symbiotic Turbellaria are simple and, so far as is known, involve only one host. The main adaptive feature lies in the tendency, already noted, to increase the size and the fecundity of the female part of the reproductive system. The increased production of eggs is obviously an advantage, and probably a necessity, in entocommensal and entoparasitic animals and is taken to extreme lengths in the trematodes and cestodes. These two classes also show asexual multiplicative stages, which in the digenetic trematodes and cestodes is closely linked with the occurrence of primary, secondary, or even tertiary hosts in the life cycle. In contrast, asexual multiplication has not evolved in symbiotic turbellarians and each adult individual normally is the sole product of one fertilized egg. This is somewhat surprising, as the potential for asexual multiplication does exist in the Turbellaria. Some free-living rhabdocoels (e.g., *Stenostomum* and *Microstomum* spp.) undergo transverse fission into pairs or chains of zooids, with each one becoming well differentiated toward the adult form before breaking free from the chain. Also, some freshwater triclads (*Dugesia, Phagocata,* and *Polycelis*) can multiply by transverse fission, and throughout the class the potential for asexual regeneration of damaged parts is common and well documented. It would be thought that the capacity to reproduce asexually within the host, even in the context of a simple life cycle, would have conferred distinct advantages in terms of the survival of the species, but there has apparently been no adequate stimulus for its evolution.

Viviparity is occasionally found among symbiotic turbellarians, but there are no signs of any further development toward polyembryony of the type seen in digenetic trematodes. Neither is there any reason to link this phenomenon with the life style as it occurs also in a few related free-living species. Thus, in the rhabdocoel family Graffillidae the free-living *Bresslauilla* spp. and the symbiotic *Paravortex cardii* and *P. gemellipara* (entocommensal in lamellibranchs) are all viviparous. In *P. cardii*, whose mature adult stage lives in the stomach of *Cardium edule*, the embryos develop to an advanced stage in the mesenchyme of the parent which then ruptures and releases them. The freed larvae pass on into the host's intestine where they grow and probably copulate (Hallez 1909; Atkins 1934). The young adults are then believed to leave the host with its feces and the life cycle is completed, presumably, when a fertilized worm is taken in on the inhalent current of a new host and passed to the stomach.

The life cycle of the ectocommensal temnocephalids has already been discussed in some detail and shown to be essentially the same as that of free-living turbellarians, except that the eggs are cemented onto the host animal. This, too, is basically the case with the ectocommensal triclads.

Nothing is known of the life cycle of the entocommensal umagillids other than the fact that they show an increased emphasis on the female part of the reproductive system and a high rate of egg production. In species living in the gut of the host, the eggs or young presumably pass out with the feces, but it is not known at what stage they hatch or how the young rhabdocoel enters and establishes itself in a new host. The eggs each bear a long whip like flagellum, which may become entangled in the encrusting algae upon which the echinoid host browses, and in this manner they increase their chances of ingestion. If the entocommensals inhabit the coelom of the host it is difficult to see how the eggs reach the exterior, unless they first migrate by some means into the gut.

Even in the parasitic Fecampiidae the life cycle remains simple. In *Kronborgia* a somewhat unusual feature is that the female produces a cocoon large enough to accommodate temporarily both herself and the male. Some free-living species, notably the triclads, also produce cocoons but these are always much smaller than the adult individual. Many thousands of eggs are laid within the cocoon and *Kronborgia*, in fact, shows the most extreme development of this tendency of the symbiotic turbellarians to increase egg production. The Fecampiidae also provide the only documented instances of symbiotic turbellarians whose young individuals gain entry to new hosts by actively penetrating the host's integument. Christensen and Kanneworff (1965) believe that the penetration mechanism is chemical in nature rather than mechanical. Therefore, this action is comparable to that shown by many trematode cercariae.

Thus, in all known cases the young stages of symbiotic Turbellaria establish themselves in new hosts by fairly direct means. No instances are known of the utilization of the host's food chain to gain entry to its body, as is commonly done by digenetic trematodes and cestodes. In this respect, then, the symbiotic Turbellaria more nearly resemble the monogenetic trematodes with their direct life cycles.

DISCUSSION

Although the symbiotic Turbellaria span the entire spectrum of types of symbioses, from temporary shelter associations through ecto- and entocommensalism to ecto- and entoparasitism, on the whole they show remarkably few departures from the typical turbellarian patterns of morphology, physiology, and life cycles. Apart from a few extreme cases, such as the adult Fecampiidae, all are quickly recognizable as turbellarians and the overall impression given is that these are species of basically free-living families which have merely extended their range of habitats to include the surfaces or internal cavities of other animals. The fact that such extensions into new habitats and the subsequent adoption of new life styles have been possible without necessitating major changes in anatomy and physiology shows the plasticity and adaptability of the basic turbellarian grade of organization. This versatility, linked with what appears to be an inherent platyhelminth trait to form associations with other species, demonstrates how the trematodes and cestodes may well have arisen either directly from turbellarian ancestors, or from a common ancestral stock which early split into the forerunners of the modern free-living and symbiotic classes.

The modern symbiotic Turbellaria as a group show the entire range of types of adaptations to symbiotic life that occur in the Trematoda and Cestoda, with the marked exception of asexual multiplicative stages in the life cycle. These adaptations are generally not taken to the extreme degree of elaboration seen in the wholly parasitic classes, but they do demonstrate that the potential for such adaptation is present in the turbellarian stock. Thus the turbellarian epidermis, the adhesive mechanisms, the alimentary system and the physiology of nutrition, and the life cycle have all been modified, although to varying extents, along the same lines as have those of the trematodes and cestodes. The reproductive system has been little modified, except for some instances of increases in the size of the ovaries and yolk glands to increase the production of eggs. This too foreshadows the situation in the parasitic classes where the process is taken further. In the parasites the oviducts or other ducts are elongated to form a uterus for storage and development of the fertilized eggs so that the embryo can hatch after laying whenever optimum conditions for its survival prevail.

It would seem then that the symbiotic turbellarians show the

basic capacity of the turbellarian stock to adapt itself to the symbiotic life style, and that since this potential is present and
widespread in the Turbellaria, the stock is in a sense preadapted
for the evolution of parasitism in the other two classes of the
phylum. In this connection, the adaptive modification of carbohydrate metabolism in the rhabdocoel family Umagillidae is of extreme
interest. It was shown earlier that in this aspect of its physiology, as shown by its increased glycogen content, the umagillid
Syndesmis resembles the trematodes and cestodes rather than the free-
living turbellarians. In this genus, as in digenetic trematodes from
aerobic habitats, environmental oxygen tension cannot be the factor
causing this emphasis on carbohydrate metabolism. It seems more
likely that it is somehow linked with increased egg production to
increase the chances of establishing the new generation in new hosts.
Thus, the extremely modified carbohydrate metabolism of the cestodes,
which manifestly is linked with the oxygen tension of the environment
and enables the parasites to survive by means of anaerobic glycolysis, may well have its origins in a simpler situation of the type
seen in the rhabdocoel *Syndesmis*. In other words, the adaptation in
carbohydrate metabolism linked with increased egg production, which
occurs in *Syndesmis* within the range of tolerance of modification of
the turbellarian grade of organization, provides the potential or
perhaps even the preadaptation for subsequent more extreme modifications along the same lines but linked with respiratory physiology.

Syndesmis also shows a trend away from the free-living turbellarians and toward the trematode and cestode type of metabolism in
the array and concentration of its free amino acid pool. The significance of this is not clear, but again this phenomenon may be demonstrating a basic adaptability in turbellarian physiology which allows
modifications for life within another animal, first at a fairly simple
level and then, with further elaboration, at a more complex and specialized one.

Other than these modifications in carbohydrate and amino acid
metabolism, the entocommensal rhabdocoels show surprisingly little
change in their physiology. The nutritional physiology is similar
to that of the ectocommensal and free-living species, largely or
entirely because the diet is similar. The ready availability of
suitable prey organisms within echinoderms enables the entocommensal
rhabdocoels to flourish within these hosts, and adopt a curious
lifestyle in which they feed as predators but which necessitates
modification of some other aspects of their biology along lines followed by parasitic species. One species of umagillid, *Syndesmis
franciscana*, shows how the nutritional physiology could then be modified toward parasitism. This species occasionally ingests by chance
host coelomocytes and digests them along with its normal food of co-
commensal ciliates. A shift in dietary preferences toward host tissues would transform *S. franciscana* into a parasitic flatworm with
a life style and physiology very like that of a digenetic trematode.
Presumably, in the abundance of prey organisms in the habitat, the
stimulus for such a shift has never occurred.

This abundance within echinoderms of the type of food favored by rhabdocoels is presumably why this phylum provides the majority of hosts for the entocommensal turbellarians. It has allowed full exploitation of the turbellarian potential for entocommensalism but, paradoxically, not stimulated the evolution of parasitism. In contrast, the establishment of symbioses with crustaceans has stimulated the evolution of this life style, although in only a few cases. The ectocommensal temnocephalids feed in the typical turbellarian fashion and appear to cause no ill effects on their hosts, but some of the entosymbionts have evolved into true parasites. Their adaptations involve loss of the gut in the adult stage, and parallel modification of the body surface for absorption of nutrients, thus demonstrating the potential for the type of adaptation to parasitic life characteristic of cestodes.

Other symbiotic turbellarians, fewer in number, have become associated with molluscs, but their precise status in these hosts is usually difficult to define. They do not show any particular adaptations so far as is known, and interest in them is attributable to the fact that they are symbiotic with hosts which are believed to be the primitive hosts of digenetic trematodes. The ancestors of the latter, however, must have had the potential for modification of the life cycle to permit the evolution of asexual multiplication, and also the potential for colonization of further host types, namely, the vertebrates, when these evolved. Some modern rhabdocoels symbiotic in molluscs are viviparous, and it is tempting to suggest that this reflects a potential for the type of development seen in the sporocysts and redia of modern digeneans. Some related free-living rhabdocoels are also viviparous, however, so this "preadaptation," if indeed it can be regarded as such, is an independent turbellarian trait and has not evolved as an adaptation to entocommensal life in molluscs.

The majority of the symbiotic turbellarians are rhabdocoels and so it would appear that this group, among all the Turbellaria, shows the strongest development of the platyhelminth trait of living in association with other animals. This lends support to the various theories, arrived at from the consideration of other types of evidence, that the turbellarian or turbellarian like ancestors of the trematodes and cestodes were rhabdocoel stock. Of the two principal modern theories, the first, summarized by Bychowsky (1957) and supported and developed by Llewellyn (1965), suggests that the cestodes arose from early monogeneans and hence only indirectly from free-living flatworms. The second, summarized by Stunkard (1962) proposes separate origins of the trematodes and cestodes, with both having arisen directly from ancestral free-living flatworms. The modern symbiotic Turbellaria, as far as they are known, do not provide particularly substantial evidence for either of these theories except that they are indeed living demonstrations of the fact that ecto- and entocommensalism and parasitism have arisen quite independently and more than once among the Turbellaria. Thus they do show that, at least in terms of life styles, Stunkard's suggestion of a diphyletic origin of the two classes is a practical possibility.

The various types of symbiotic relationships seen in the modern Turbellaria probably represent, in most cases if not all, end points in the development of these particular symbioses. Some are known to be of very ancient origin and yet have still not evolved to a climax of parasitism. The best example here is the temnocephalid-crustacean symbiosis, where the geographical distribution dates the relationship's origin as being at least before the middle of the Tertiary period and possibly as early as the Cretaceous (p. 129). Despite their early origin, the temnocephalids as a group have remained ectocommensals and species from habitats as far apart as Central America and New Zealand show no significant differences in their way of life or their nutritional physiology.

Nevertheless, the different symbioses that have become established in the Turbellaria do indicate possible pathways that might have been followed by the Trematoda and Cestoda during their early evolution from the supposed free-living ancestors. Association with appropriate hosts, at first by chance or in search of temporary shelter, could have led either directly or via ecto- and entocommensalism to the present life style. The nature of the hosts in these early stages was probably all-important, with those species associating with molluscs and invertebrates other than echinoderms being the ones to give rise eventually to the modern trematodes and cestodes.

Once these early symbioses had become established, the progressive development and elaboration of the turbellarian potential for adaptive modifications in the epidermis, adhesive organs, reproductive system, life cycle, and nutritional physiology would then complete the transformation of the predatory free-living ancestor into the modern parasitic trematodes and cestodes.

SUMMARY

The Turbellaria show a number of features in common with the parasitic classes of the Platyhelminths, consistent with the theory that the phylum arose from turbellarian or turbellarian like ancestors, and these include a tendency to form symbioses with higher organisms. The turbellarians are predominantly free-living predators, but over 120 species, from 27 families representative of all the major subdivisions of the class, live symbiotically with a variety of hosts. The majority are symbiotic with echinoderms and crustaceans, but molluscs, annelids, a few other invertebrates, and, in three species only, vertebrates are also used as hosts. The types of symbioses span a very broad spectrum from simple shelter associations, which demonstrate the possible origins of the symbiotic habit, through facultative and obligate ectocommensalism to entocommensalism and ecto- and entoparasitism. It is suggested, from a consideration of various aspects of these symbioses, that the original shelter-seeking factor was quickly reinforced by a nutritional one during the establishment of the various symbioses, and that either the availability in the new habitat of the kind of organisms upon which free-living turbellarians normally prey, or the nature of the host's feeding mechanism, or both, were the important components of this nutritional factor. The majority of the ecto- and entocommensal turbellarians feed in the same way and generally on the same types of food as the free-living species. The predominance of echinoderms as hosts for the entocommensals is ascribed to the common occurrence in these animals of co-commensals, such as ciliate protozoa, upon which the symbiotic flatworms feed. Only a few species have become entoparasitic; most of these have lost the gut and presumably feed like cestodes.

The symbiotic Turbellaria show surprisingly few modifications of the basic turbellarian patterns in their morphology, physiology, and life history. When modifications do occur, however, they foreshadow the fairly extreme modifications seen in trematodes and cestodes. Thus, such modifications as do occur are found in the epidermis, the adhesive organs, the alimentary system, and the reproductive system. Two interesting physiological modifications occur in an entocommensal rhabdocoel, in which both the carbohydrate metabolism and free amino acid pool are of the type characteristic of trematodes and cestodes rather than of free-living turbellarians. The life cycles in all cases remain simple, although there is a tendency toward increased egg production, and there are no secondary hosts or asexual multiplicative stages.

The variety of symbioses found in the Turbellaria indicate a number of potential pathways to parasitism which may have been followed by trematodes and cestodes during their evolution from their free-living ancestors. The turbellarian stock is seen to possess the potential to evolve morphological and physiological features characteristic of these wholly parasitic classes. On the whole the turbellarian symbioses themselves represent end points in the development of those particular types of association; some are of very ancient origin and do not appear to have changed in nature over many thousands of years.

LITERATURE CITED

Atkins, D. 1934. Two parasites of the common cockle *Cardium edule;* a rhabdocoele *Paravortex cardii* Hallez and a copepod *Paranthessius rostratus* (Canu). J. Mar. Biol. Ass. U. K. 19:669-76.

Averinzev, S. 1925. Über eine neue Art von parasitaren Tricladen (*Micropharynx*). Zool. Anz. 64:81-84.

Baer, J. G. 1931. Étude monographique du groupe des Temnocéphales. Bull. Biol. Fr. Belg. 55:1-57.

_____. 1951. Ecology of animal parasites. Urbana: Univ. of Illinois Press.

Bashiruddin, M., and Karling, T. G. 1970. A new entocommensal turbellarian (Fam, Pterastericolidae) from the sea-star *Astropecten irregularis*. Z. Morph. Tiere 67:16-28.

Baylis, H. A. 1949. *Fecampia spiralis,* A cocoon–forming parasite of the antarctic isopod *Serolis schytei.* Proc. Linn. Soc. London 161:64–71.

Beauchamp, P. de. 1921. Sur quelques rhabdocoeles des environs de Dijon. Congr. Strasbourg: C. R. Assoc. franc. Av. Scie.

_____. 1961. Classe des Turbellariés. In Traité de Zoologie, ed. P. –P. Grassé, Vol. 4. Paris: Masson et Cie.

Bock, S. 1925. Papers from Dr. Th. Mortensen's Pacific Expedition 1914–16. XXV. Planarians. Parts I–III. Vidensk. Meddr dansk naturh. Foren. 19:1–84, 97–184.

_____. 1926. Eine Polyclade mit muskulösen Drüsenorganen rings um den Körper. Zool. Anz. 66:133–38.

Brand, T. von. 1936. Studies on the carbohydrate metabolism in Planarians. Physiol Zool. 9:530–41.

_____. 1966. Biochemistry of parasites. New York: Academic Press.

Bresciani, J., and Køie, M. 1970. On the ultrastructure of the epidermis of the adult female of *Kronborgia amphipodicola* Christensen and Kanneworff, 1964 (Turbellaria, Neorhabdocoela). Ophelia 8:209–30.

Bychowsky, B. E. 1957. Monogenetic Trematodes, their classification and phylogeny. Academy of Sciences, U.S.S.R., Moscow and Leningrad (in Russian). Translated by W. J. Hargis and P. C. Oustinoff. Washington: Ann. Inst. Biol. Sci.

Caullery, M., and Mesnil, F. 1903. Recherches sur les *Fecampia* Giard, Turbellariés Rhabdocoeles, parasites internes des Crustacés. Ann. Fac. Sci. Marseille 13:131–68.

Christensen, A. M., and Kanneworff, B. 1965. Life history and biology of *Kronborgia amphipodicola* Christensen and Kanneworff (Turbellaria, Neorhbadocoela). Ophelia 2:237–51.

Dakin, W. 1912. *Buccinum.* Proc. Trans. Liverpool Biol. Soc., Vol. 26, Mem. No. 2.

Dörler, A. 1900. Neue und wenig bekannte rhabdocöle Turbellarien. Z. wiss. Zool. 68:1–42.

Fernando, W. 1945. The storage of glycogen in the Temnocephaloidea. J. Parasitol. 31:185–90.

Fyfe, M. F. 1942. The anatomy and systematic position of *Temnocephala novae-zealandiae* Haswell. Trans. Roy. Soc. N. Z. 72: 253–67.

Giard, M. A. 1886. Sur une rhabdocoele nouveau, parasite et nidulant (*Fecampia erythrocephala*). C. R. Acad. Sci. Paris 103: 499–501.

Gibson, R., and Jennings, J. B. 1969. Observations on the diet, feeding mechanisms, digestion and food reserves of the entocommensal rhynchocoelan *Malacobdella grossa.* J. Mar. Biol. Ass. U. K. 49:17–32.

Girard, C. 1850. Two marine species of Planariae. Proc. Boston Soc. Nat. Hist. 3:264.

Gonzales, M. D. P. 1949. Sobre a digestão e repiração des Temnocephalas (*Temnocephalus bresslaui* spec. nov.). Bol. Fac. Filos. Ciênc. Univ. São Paulo (Zool.). 14:277–323.

Graff, L. von. 1904-1908. Acoela und Rhabdocoelida. In Klassen und Ordnungen des Tier-Reichs, ed. H. G. Bron, 4, 1.1:1-2599.

Hallez, P. 1909. Biologie, organization, histologie et embryologie d'un rhabdocoele parasite du *Cardium edule* L., *Paravortex cardii* n. sp. Archs. Zool. exp. gén., Ser. 4, 9:1047-49.

Halton, D. W. 1966. Occurrence of microvilli-like structures in the gut of digenetic trematodes. Experientia 22:828-29.

_____. 1967. Studies on glycogen deposition in Trematoda. Comp. Biochem. Physiol. 23:113-20.

Hargis, W. J. 1955. Host specificity of monogenetic trematodes. Assoc. Southeastern Biol. Bull. 2:6.

_____. 1957. The host specificity of monogenetic trematodes. Exp. Parasitol. 6:610-25.

Hickman, V. V. 1956. Parasitic turbellaria from Tasmanian Echinoidea. Papers Proc. Roy. Soc. Tasmania 90:169-81.

_____. 1967. Tasmanian Temnocephalidae. Papers Proc. Roy. Soc. Tasmania 101:227-50.

_____, and Olsen, A. M. 1955. A new turbellarian parasite in the sea-star *Coscinasteris calamaria* Gray. Papers Proc. Roy. Soc. Tasmania 89:55-63.

Hyman, L. H. 1951. The Invertebrates: Vol. 2, Platyhelminths and Rhynchocoela. New York: McGraw-Hill.

_____. 1955. The Invertebrates: Vol. 4, Echinodermata. New York: McGraw-Hill.

_____. 1967. The Invertebrates: Vol. 6, Mollusca. New York: McGraw-Hill.

Jägerskiöld, L. A. 1896. Ueber *Micropharynx parasitica*, n.g., n. sp., eine ectoparasitische Triclade. Ofv. Vet. Akad. Förhandl. Stockholm 53:707-15.

Jägersten, G. 1942. Zur Kenntnis von *Glanduloderma myzostomatis* n. gen., n. sp., einer eigentümlichen, in Mysoztomiden schmarotzenden Turbellarienform. Ark. Zool. 33(3):1-24.

Jennings, J. B. 1957. Studies on feeding, digestion and food storage in free-living flatworms. Biol. Bull. mar. biol. Lab. Woods Hole 112:63-80.

_____. 1968a. Platyhelminthes: Nutrition. In Chemical zoology, ed. M. Florkin and B. T. Scheer, pp. 303-26. New York:Academic Press.

_____. 1968b. Feeding, digestion and food storage in two species of temnocephalid flatworms (Turbellaria: Rhabdocoela). J. Zool. London 156:1-8.

_____. 1968c. A new temnocephalid flatworm from Costa Rica. J. Nat. Hist. 2:117-20.

_____. 1971. Parasitism and commensalism in the Turbellaria. In Advances in parasitology, Vol. 9, ed. B. Dawes, pp. 1-32. New York: Academic Press.

_____. 1972. Feeding, digestion and assimilation in animals. 2nd ed. London: Macmillan.

_____. 1973. Digestive physiology of the Turbellaria. In Biology of the Turbellaria, ed. N. W. Riser and M. Patricia Morse.

Libbie H. Hyman Memorial Volume. New York: McGraw-Hill. (In press.)

_____, and Mettrick, D. F. 1968. Observations on the ecology, morphology and nutrition of the rhabdocoel turbellarian *Syndesmis fransiscana* (Lehmann, 1946) in Jamaica. Carib. J. Sci. 8(1-2): 57-69.

Jhering, H. von. 1880. *Graffila muricola*, eine parasitische Rhabdocoele. Z. wiss. Zool. 34:147-74.

Kaburaki, T. 1922. On some Japanese Tricladida Maricola, with a note on the classification of the group. J. Coll. Sci. Imp. Univ. Tokyo 44:1-54.

Karling, T. G. 1970. On *Pterastericola fedotovi* (Turbellaria) commensal in sea stars. Z. Morph. Tiere 67:29-39.

Kato, K. 1935a. *Stylochoplana parasitica* sp. nov., a polyclad parasite in the pallial groove of the Chiton. Annotnes Zool. Jap. 15:123-29.

_____. 1935b. *Discoplana takewakii* sp. nov., a polyclad parasite in the genital bursa of the Ophiuran. Annotnes Zool. Jap. 15: 149-56.

Laidlaw, F. 1902. *Typhlorhynchus nanus*. Quart. J. Microscop. Sci. 45:637-52.

Llewellyn, J. 1965. The evolution of parasitic Platyhelminthes. In Evolution of parasites, ed. Angela E. R. Taylor, pp. 47-78 Third Symposium of the British Society for Parasitology. Oxford: Blackwell.

Marcus, E. 1952. Turbellaria Brasileiros (10). Bol. Fac. Fil. Ciênc. Letr. Univ. Sao Paulo, Zool. 17:5-188.

Matjasic, J. 1957. Biologie und Zoogeographie der europäischen Temnocephaliden. Zool. Anz. Suppl. 21:477-82.

Mettrick, D. F., and Boddington, M. J. 1972a. The chemical composition of some marine and freshwater turbellarians. Carib. J. Sci. 12(1-2):1-7.

_____. 1972b. Amino acid pools of *Syndesmis franciscana* (Turbellaria: Platyhelminthes) and of host coelomic fluid. Can. J. Zool. 50:411-13.

_____, and Jennings, J. B. 1969. Nutrition and chemical composition of the rhabdocoel turbellarian *Syndesmis franciscana*, with notes on the taxonomy of *S. antillarum*. J. Fish. Res. Bd. Canada 26: 2669-79.

Mrazeck, A. 1906. Ein europäischer Vertreter der Gruppe Temnocephaloidea. Sber. K. böhm. Ges. Wiss. 36:1-7.

Naylor, E. 1952. On *Plagiostomum oyense* de Beauchamp, an epizoic turbellarian new to the British Fauna. Ann. Rep. Mar. Biol. Sta. Port Erin 64:15-21

_____. 1955. The seasonal abundance on *Idotea* of the cocoons of the flatworm *Plagiostomum oyense* de Beauchamp. Ann. Rep. Mar. Biol. Sta. Port Erin 67:25-30

Pearse, A. S., and Wharton, G. W. 1938. The oyster "leech" *Stylochus inimicus* Palomli, associated with oysters on the coasts of Florida. Ecol. Monogr. (Durham, North Carolina) 8:605-55.

Prudhoe, S. 1968. A new polyclad turbellarian associating with a hermit crab in the Hawaiian Islands. Pacif. Sci. 22:408-11.

Read, C. P. 1970. Parasitism and symbiology. New York: The Ronald Press.

Reisinger, E. 1930. Zum ductus genito-intestinalis Problem. 1. Über primäre Geschlechtstraktdarmverbindungen bei rhabdocoelen Turbellarien. Z. Morph. Ökol. Tiere 16:49-73.

Schaefer, C. W. 1971. Observations on temnocephalid hosts and distributions. Z. f. zool. Systematik u. Evolutionsforschung 9: 139-43.

Smyth, J. D. 1966. The physiology of trematodes. Edinburgh: Oliver and Boyd.

_____. 1969. The physiology of cestodes. Edinburgh: Oliver and Boyd.

Sokal, R. R., and Rohlf, F. J. 1969. Biometry. San Francisco: W. H. Freeman and Co.

Stunkard, H. W. 1962. The organisation, ontogeny and orientation of the Cestoda. Quart. Rev. Biol. 37:23-34.

Syriamiatnikova, I. P. 1949. A new turbellarian parasite of fish, *Icthyophaga subcutanea* n. g. nov. sp. (in Russian). C. R. Acad. Sci. Moscow 68(2):805-808

Ugolev, A. M. 1960. The existence of parietal contact digestion. Bull. exp. Biol. and Med. (U.S.S.R.) 49:1-12.

_____. 1965. Membrane (contact) digestion. Physiol.Rev. 45:555-95.

Westblad, E. 1937. Die Turbellarien-Gattung *Nemertoderma* Steinböck. Acta Soc. Fauna Flora fenn. 60:45-89.

_____. 1949. On *Meara stichopi* (Bock) Westblad, a new representative of Turbellaria archoophora. Ark. Zool. 1:43-57.

_____. 1953. New Turbellaria parasites in echinoderms. Ark. Zool. 5:269-88.

_____. 1955. Marine Alloeocoels (Turbellaria) from N. Atlantic and Mediterranean coasts. Ark Zool. 7:491-526.

Wheeler, W. M. 1894. *Syncoelidium pellucidum*, a new marine triclad. J. Morphol. 9:167-94.

Wilhelmi, J. 1909. Tricladen. Fauna und Flora des Golfes von Neapel u. d. angr. Meeresabschnitte. Monographie Berlin 32: 1-405.

Metabolic pattern of a trematode and its host: a study in the evolution of physiological responses

W. B. Vernberg and F. J. Vernberg
Belle W. Baruch Coastal Research Institute
University of South Carolina
Columbia, South Carolina

To survive, all animals must live in equilibrium with an environment that includes both abiotic and biotic factors, and the ability of organisms to survive in a particular environmental complex is the result of evolutionary changes over an undetermined period of time. The successful organism is one which has adaptive strategies enabling it not only to survive, but also to exploit changing environmental regimes. Although many marine organisms occupy diverse habitats at different stages in their life cycle, perhaps none encounters a wider range of environmental complexes than do digenetic trematodes. Hence, digenetic trematodes are excellent experimental animals to use for the purpose of studying the general evolutionary problem of physiological adaptation to environmental stress with special emphasis on analyzing the interrelationship between nongenetic environmentally induced (phenotypic) adaptations and genotypic adaptations.

The different environments which the various stages of a trematode encounter are represented in over simplified form in Figure 1. Such habitat diversity poses many intriguing puzzles, such as what physiological strategies the various stages of a trematode have evolved enabling it to cope with different environments, how much each host imprints its own physiological characteristics on those of the fluke, and what physiological responses the flukes have evolved independently of those of the host animals.

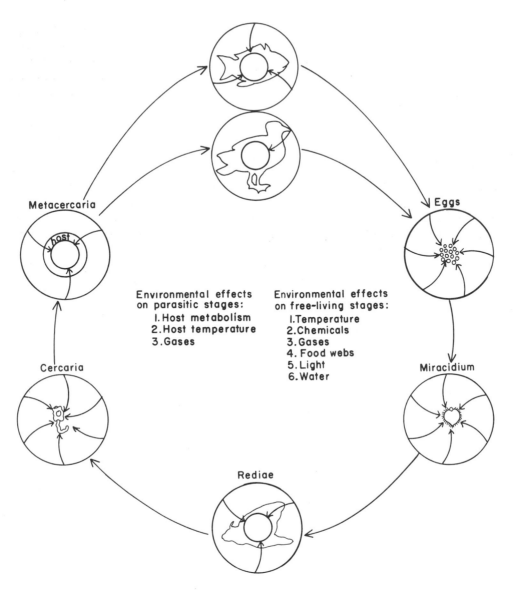

Figure 1. A simplified scheme of the different environments encountered by the various stages in the life cycle of a
trematode.

LARVAL ADAPTATIONS

Since trematodes live in extremely diverse ecological situations
throughout their life cycle, it is not surprising that in some aspects
the animals show striking physiological adaptations to particular sets
of environmental conditions. For example, some stages in the life
cycle seem to have distinctly different metabolic requirements than
others do, which can be correlated with the environmental complex

that this stage encounters. The need for metabolic oxygen, for example, varies greatly with the stage in the life cycle. Adults are quite resistant to oxygen lack, whereas the free-living stages can tolerate anaerobiosis for only brief periods of time (Vernberg 1969). Metabolic responses to low oxygen tensions also illustrate differences in physiological adaptations by the different stages. The rediae of *Himasthla quissetensis*, which occur in the digestive gland of the mud-flat snail, *Nassarius obsoleta*, metabolically are relatively independent of oxygen tension, but the metabolic rate of the free-living cercarial stage is completely dependent on oxygen tension (Vernberg 1963). These responses can readily be correlated with the environmental conditions each species encounters in nature: the free-living cercarial stage has access to a greater oxygen supply compared to that available to a parasitic stage living in the digestive tract or tissue of another organism.

Many aspects of the metabolism of the free-living cercarial stage can be interpreted in light of the same factors that regulate the metabolism of other free-living planktonic forms. Actively swimming cercariae and other zooplankton tend to require more oxygen than do less active species. There is also a correlation between body size and oxygen uptake rates in these forms when measured under normal environmental conditions. Interspecific comparisons of metabolism and body size (body N) were calculated for five species of cercariae, all of which developed in the same first intermediate host, *N. obsoleta*. The formula $O_2 = aW^b$ or $\log O_2 = \log \underline{a} + \underline{b} \log W$ was used. In this equation,

O_2 = oxygen consumed per unit time and unit weight
W = weight of animal, expressed for these cercariae as body N
a and b = coefficients, in which $\underline{a}$ represents the intercept of the y-axis and $\underline{b}$ represents the slope of the function in the logarithmic plot.

The relationship of oxygen consumption to body size can be determined from the slope of the line (the constant $\underline{b}$). If $\underline{b}$ is 1, then the rate of oxygen utilization is significantly proportional to weight. The correlation between size and oxygen consumption decreases as the $\underline{b}$ value decreases (Vernberg and Hunter 1959). For the five species of cercariae, the $\underline{b}$ value was 0.79, which is comparable to values found for other species of planktonic larvae (Zeuthen 1947) (Fig. 2).

However, cercariae do not always respond to environmental fluctuations in the same manner as other free-living zooplankton. Most poikilothermic animals can survive temperatures slightly in excess of the range normally encountered in nature. In temperate zone animals the upper range of thermal tolerance is generally from approximately 33 to 40C. A comparison of the thermal responses of cercariae with other free-living marine organisms shows that the limit of thermal tolerance is indeed quite different from that of other organisms living in essentially the same thermal environment (Fig. 3). Thus, larvae of species of the common fiddler crab, *Uca*, can tolerate temperatures up to 40C, whereas the upper thermal limit of the cercariae of

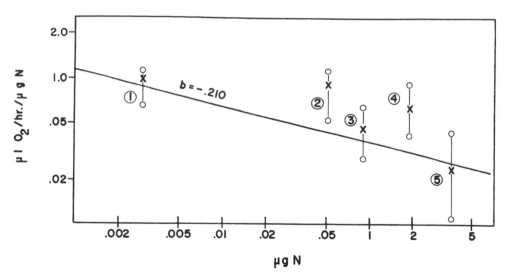

Figure 2. Relationship of oxygen consumption to body N in five
species of cercariae. (1) *Gynaecotyla adunca*, (2) *Cardiocephalus brandesii*, (3) *Zoögonus lasius*, (4) *Lepocreadium
setiferoides*, and (5) *Himasthla quissetensis*. (0-0 =
range of Q_{O2} obtained for each species.)(From Vernberg and
Hunter 1959.)

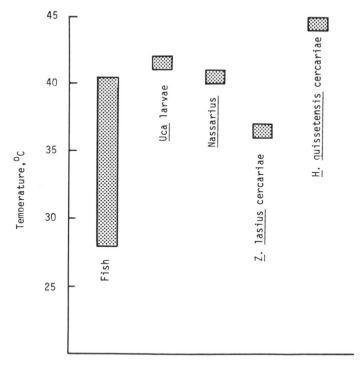

Figure 3. Upper thermal tolerances of cercariae in comparison to
other free-living marine organisms.

Zoögonus lasius is 36C. On the other hand, the cercariae of *H. quis-setensis* has an upper thermal limit of 44C, a temperature well above that of most free-living temperate zone zooplankton (Vernberg 1961). The upper thermal limit of *N. obsoleta*, which serves as the first intermediate host for both trematode species, is between that of these two species of cercariae. It is quite obvious that the upper thermal limits of the cercariae are linked not to their free-living environmental envelope, but rather to the thermal environment they will encounter in their adult host: *H. quissetensis* utilizes a bird host whose body temperature is approximately 41C; the body tempera-ture of the fish host of *Z. lasius* varies but normally does not exceed 32C.

Within the organism the ultimate seat of control and regulation of life processes is found in the genetic material of the cell. Col-lectively, the sum total of all genes carried by an organism is the genotype and the genotype is constant throughout the diploid phases of the life cycle, except if modified by a mutation. The expression of the genes in terms of function at any one point in time is called the phenotype. Thus, the phenotypic response of the animal to a given environmental situation is determined by its genotype. It is difficult if not impossible to obtain the exact genotypic limits of an organism, but an approximation of the genotype of thermal response can be gained by measuring the metabolic-temperature response of parasites under cold- and warm-acclimated conditions. From compari-son of such responses of the intramolluscan sporocyst stage of *Z. lasius*, which utilizes fish as its adult host, and the intramolluscan redial stage of *H. quissetensis*, which utilizes birds as its adult host, it is obvious that the genotypic responses are quite different for the two species (Fig. 4). *H. quissetensis* is considerably more labile metabolically than is *Z. lasius*. The genotypic limits of response of both parasites are fairly great at the lower temperatures, but while this lability is observed at the higher temperature in the bird parasite, in *Z. lasius* it is markedly restricted at tempera-tures above 30C. This response, of course, can be correlated again with the upper lethal thermal limits of the adult host. In contrast to these interspecific differences between larval stages, a compari-son of the genotypic limits of *Z. lasius* sporocysts and both the first intermediate host, *N. obsoleta*, and the definitive fish host, the toadfish, *Opsanus tau*, clearly shows that all three have similar geno-typic responses over the temperature range they would encounter in nature (Fig. 5): limited lability at high temperatures but consider-able lability at intermediate temperatures. Metabolic-temperature responses of temperate zone birds are such that body temperatures remain constantly high except under extreme thermal stress. Thus, a bird parasite must be able to tolerate a constant high tempera-ture as an adult, but also must be able to adapt to the lower temper-atures it encounters during larval development. The genotype of *H. quissetensis* reflects the wide thermal range it will encounter during its lifetime. Similar genotypic lability patterns were reported for tropical and temperate zone fiddler crabs (genus *Uca*).

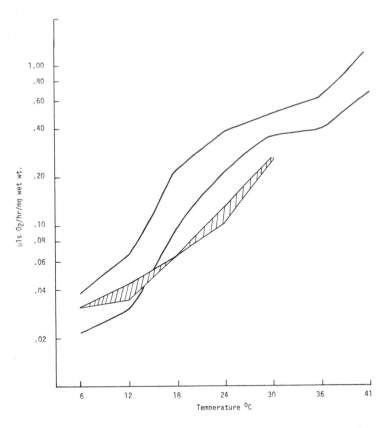

Figure 4. The "genotypic" limits of metabolic thermal response of the intramolluscan larvae of the bird parasite, *H. quissetensis*, and the fish parasite, *Z. lasius* (based on data from Vernberg and Vernberg 1965).

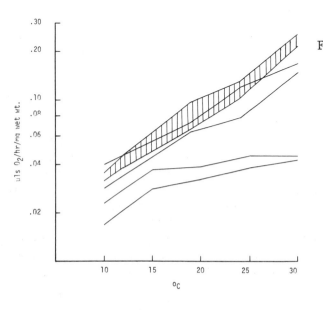

Figure 5. The "genotypic" limits of metabolic thermal response of the larvae of *Z. lasius* and its intermediate host (*N. obsoleta* and its definitive host (*O. tau*). Limits for *Z. lasius* and *N. obsoleta* are based on data from Vernberg and Vernberg 1965.

The warm-water species, *U. rapax*, shows little lability in its meta-bolic-temperature response at low temperatures but great lability at higher temperatures. In contrast, the temperate zone species, *U. pugnax*, exhibited the greatest degree of lability at low temperature. Thus, organisms appear to be genotypically restricted and unable to acclimate at environmental extremes (Vernberg 1959).

One question that has long been asked and is as yet unanswered is why some animals can survive temperature extremes better than others. Certainly the differences in thermal tolerance between lar-val trematodes sharing a common larval environment offer a case in point. It has been suggested that the cause of thermal death is in some way related to lipid metabolism, perhaps to the biophysical changes in cellular lipids (Somero and DeVries 1967; Caldwell and Vernberg 1970). Two enzymes, glucose-6-phosphate dehydrogenase (G-6P) and 6-phosphogluconic dehydrogenase (6PG), are important in lipid synthesis. Both G-6P and 6PG have been reported in cercariae and adult *Shistosoma mansoni*. Coles (1970) reported similar electro-phoretic migration patterns of G-6P dehydrogenases from daughter sporocysts and adults of *S. mansoni*, whereas other enzymatic migra-tion patterns of larval and adult stages are often quite dissimilar. He stated in this paper that"it would be interesting to know how far the similar migration of glucose-6-phosphate dehydrogenase from the daughter sporocysts and the adult worms represents a similar metabol-ism adapted to a parasitic way of life in the two different hosts." Perhaps the similar migration patterns are not connected with the parasitic way of life as such, but instead with the thermal tolerance of the flukes.

The activity of G-6P and 6PG dehydrogenase has been assayed spectrophotometrically in the intramolluscan stages of *Z. lasius* and *H. quissetensis*. Enzymatic activity levels in the larvae of the parasite utilizing a fish definitive host (*Z. lasius*) are approxi-mately half those of the larvae utilizing a bird definitive host (*H. quissetensis*)(Fig. 6). Since these two enzymes are linked to lipid synthesis, and if temperature tolerances of the parasites are indeed a function of biophysical changes in cellular lipids, then it is tempting to speculate that these marked differences in enzymatic activity between the fish and bird parasites are in some way related to their thermal tolerances.

Another aspect in which the response of the cercariae does not appear to be correlated with the environmental conditions of either the first intermediate host or its free-living environment is the response to salinity. Euryhalinity, or tolerance to fluctuations in salinity, is a trademark of estuarine animals. In a study[*] on salin-ity tolerances of *N. obsoleta* and two species of cercariae, *L. seti-feroides* and *H. quissetensis*, this mollusc was utilized as the first intermediate host. Kasschau found that South Carolina populations of *N. obsoleta* cannot tolerate salinities below 12 o/oo. This gas-tropod is generally found in relatively high salinity areas. The

---

[*]Kasschau: personal communication.

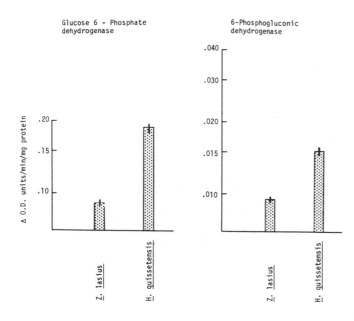

Figure 6
Enzymatic activity of G-6P and 6PG in the intramolluscan larvae of *Z. lasius* and *H. quissetensis*. Vertical bars indicate one standard error.

two species of cercariae, however, have a much wider tolerance to low salinity than does the host species (Fig. 7). The cercariae of *H. quissetensis*, for example, were able to tolerate salinities down to 3 o/oo; this is an extremely low salinity even for the most tolerant of estuarine animals. The cercariae of the fish parasite, *L. setiferoides*, also tolerated low salinity, although they were more sensitive to low salinities than cercariae of *H. quissetensis*. It

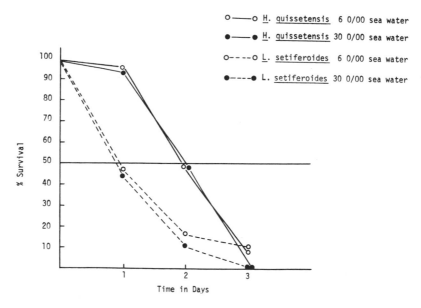

Figure 7. Survival of cercariae of *L. setiferoides* and *H. quissetensis* in 6 o/oo and 30 o/oo seawater (Kasschau, personal communication).

is of interest that shedding rates of cercariae of another trematode species, *Austrobilherzia variglandis*, were markedly depressed at salinities below 15 o/oo when *N. obsoleta* was utilized as a first intermediate host (Sinderman 1960).

Since the host species cannot tolerate low salinity waters and shedding of cercariae is depressed when salinities are low, the adaptive values of the extreme salinity tolerance of these larvae are difficult to understand. The well-developed osmoregulatory capabilities of the trematode larvae cannot be linked to the environment they meet as larvae. However, it may reflect the osmotic environment of the adult fluke. The osmotic concentration found in either the fish gut or in the bird intestine normally would be considerably less than that encountered in seawater. We know, for example, that neither fish nor bird adult parasites can survive for any length of time in full strength seawater; they will, however, survive relatively long periods in 10 o/oo.

Generally, if adult and larval stages have dissimilar environmental regimes, there are distinct differences between their tolerance levels. For example, adult fiddler crabs (genus *Uca*) from the temperate zone can tolerate greater salinity and temperature ranges than the larvae, and these differences can be correlated with the habitat temperature-salinity conditions each normally encounters (Vernberg and Vernberg 1967; Vernberg and Vernberg 1972). It is not possible to find such correlations with environmental regimes in the temperature-salinity tolerances of larval and adult trematodes. How, then, can the similar thermal and salinity responses of larval and adult trematodes be explained?

An interesting concept of the evolution of the complex life cycles has been postulated by Istock (1967). This hypothesis suggests that initially one part of the life cycle is transferred from the environmental resources at hand to a new set of supplementary resources. Two conditions must be met if this transfer is to occur: (1) the organism must have enough genetic lability - or genetic pre-adaptation - to enable it to utilize a new set of supplementary resources; and (2) there must be a net energy gain to the species with the transfer to the supplemental resources. Then each of the ecologically distinct phases in the life cycle evolves more or less independently of the others once the transfer has been made.

Stunkard (1953) has speculated that parasitism evolves through gradual and progressive adaptation, and that the organism must have undergone either physiological or morphological preadaptation or perhaps both. If, as some have suggested (Baylis 1938; Cable 1973), a stage resembling the cercaria was the adult stage, then the responses of larval trematodes to temperature and salinity could be interpreted as preadaptive ones that permitted the larvae to utilize a new resource.

ADULT ADAPTATIONS

Thermal tolerances of adult parasites agree very closely with that of their respective hosts. Parasites of fish die at lower temperatures than do turtle parasites, and turtle parasites in turn cannot survive as high a temperature as can bird parasites (Vernberg and Hunter 1961).

Once the parasite has reached its definitive host, they may select any one of a number of places in which to complete development. Does the position or location in the final host influence the metabolic rate of the animals living there? The answer is probably a qualified yes. Metabolic measurements were made on five species of trematodes living in the herring gull, *Larus argentatus*. One species, *Renicola* sp., lives in the kidney, *Austrobilharzia variglandis* in the blood vessels of the mesenteries, and three parasites, *Gynaecotyla adunca*, *H. quissetensis*, and *Cardiocephalus brandesii*, in the intestinal tract. The metabolic rate of *G. adunca* was determined in Cartesian diver respirometers. Warburg manometric techniques were used for the other species. To compare metabolic rates interspecifically it is necessary to take into account body size; this can be done by plotting the linear regression of oxygen uptake versus body weight (Fig. 8). For these five species, the <u>b</u> value was 0.65, indicating a significant correlation between body size and metabolism. Therefore, it is not possible to make a comparison between species that differ markedly in size. One pair of species, however, are the same size, *Renicola* sp. and *A. variglandis*. The former averaged 0.34 mg per worm, the latter 0.32 mg. There is a marked difference

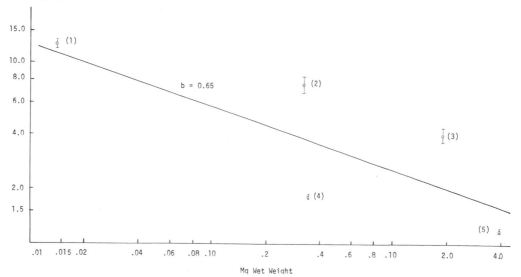

Figure 8. Relationship of oxygen consumption to body weight in five species of adult trematodes found in the herring gull, *Larus argentatus*. (1) *G. adunca*, (2) *A. variglandis*, (3) *H. quissetensis*, (4) *Renicola* sp., and (5) *C. brandesii*. Vertical bars indicate one standard error.

in the metabolic rate of these two species; the metabolic rate of
the blood parasite is approximately four times that of the kidney
parasite. Among free-living aquatic animals, those that live in
well oxygenated environments tend to have higher metabolic rates than
those living in less well oxygenated waters. Similarly, the blood-
stream habitat of *A. variglandis* has a higher oxygen level than does
the kidney where *Renicola* sp. lives, and the metabolic rates of the
parasites reflect the oxygen levels of these diverse habitats. Thus,
although metabolic rates of adult worms are generally size-dependent,
when flukes of the same size are compared, metabolic rates can be
correlated with habitat.

SUMMARY

What generalizations can be made about the physiological adaptations of digenetic trematodes throughout their complex life cycles?
In some aspects each stage has apparently evolved more or less independently of the next. Each must be able to utilize the energy
resources provided, and at the same time must be adapted to withstand
the environmental fluctuations associated with a particular way of
life. Yet, throughout the life cycle the species has a thread of
preadaptive physiological responses, a thread that is necessary for
the species to successfully make the transition from one host to
another.

LITERATURE CITED

Baylis, H. A. 1938. Helminths and evolution. In Evolution, ed.
    G. R. deBun, pp. 249-70. London: Oxford U. Press.
Cable, R. M. 1973. Phylogeny and taxonomy of trematodes with refer-
    ence to marine species. In Symbiosis in the sea, ed. W. B.
    Vernberg. U. of South Carolina Press (In press.)
Caldwell, R. S., and Vernberg, F. J. 1970. The influence of accli-
    mation temperature on the lipid composition of fish gill mito-
    chondria. Comp. Biochem. Physiol. 34:179-92.
Coles, G. C. 1970. A comparison of some isoenzymes of *Schistosoma
    mansoni* and *Shistosoma hematobium*. Comp. Biochem. Physiol. 33:
    549-58.
Istock, A. 1967. The evolution of complex life cycle phenomena:
    An ecological perspective. Evolution 21:592-605.
Sinderman, C. J. 1960. Ecological studies of marine dermatitis
    producing Shistosome larvae in northern New England. Ecology
    41:678-84.
Somero, G. N., and DeVries, A. L. 1967. Temperature tolerances of
    some Antarctic fishes. Science 156:257-58.
Stunkard, H. W. 1953. Life histories and systematics of parasitic
    worms. System.Zool. 2:7-18.

Vernberg, F. J. 1959. Studies on the physiological variation between tropical and temperate zone fiddler crabs of the genus *Uca*. Oxygen consumption of whole organisms. Biol. Bull. 117: 163–84.

_____, and Vernberg, W. B. 1967. Studies on the physiological variation between tropical and temperate zone fiddler crabs of the genus *Uca*. IX. Thermal lethal limits of southern hemisphere *Uca* crabs. Oikos 18:118–23.

Vernberg, W. B. 1961. Studies on oxygen consumption in digenetic trematodes. VI. The influence of temperature on larval trematodes. Exptl. Parasitol. 11:270–75.

_____. 1963. Respiration of digenetic trematodes. Ann. N. Y. Acad Sci. 113:261–71.

_____. 1969. Adaptations of host and symbionts in the intertidal zone. Amer. Zool. 9:357–65.

_____, and Hunter, W. S. 1959. Studies on oxygen consumption in digenetic trematodes. III. The relationship of body nitrogen to oxygen uptake. Exptl. Parasitol. 8:76–82.

_____, and _____. 1961. Studies on oxygen consumption in digenetic trematodes. V. The influence of temperature on three species of adult trematodes. Exptl. Parasitol. 11:34–38.

_____, and Vernberg, F. J. 1965. Interrelationships between parasites and their hosts. I. Comparative metabolic patterns of thermal acclimation of larval trematodes with that of their host. Comp. Biochem. Physiol. 14:557–66.

_____, and _____. 1972. Environmental physiology of marine animals. New York: Springer-Verlag.

Zeuthen, E. 1947. Body size and metabolic rate in the animal kingdom with special regard to the marine micro-fauna. Compt. rend. trav. lab. Carlsberg, Ser. Chim. 26:17–161.

Phylogeny and taxonomy of trematodes with reference to marine species

R. M. Cable
Department of Biological Sciences
Purdue University
Lafayette, Indiana

Inquiry into the phylogeny of any group of organisms is speculative and especially so when they have become highly specialized
without leaving fossil records as is true of the parasitic flatworms.
In that group, species with an intestine but lacking epidermal cilia
as adults are usually regarded as comprising the class Trematoda
with three subclasses: Monogenea, Aspidobothrea, and Digenea. Nevertheless, it has become clear that aspidobothreans and digeneans have
far more in common than either subclass has with the monogeneans.
Hence, taxonomy that reflects phylogeny would set the monogeneans
apart from the other two groups.

Although the need for such separation is generally recognized,
opinions differ as to how wide it should be. At one extreme is the
view of Bychowsky (1937, 1957), who removed the monogeneans from the
Trematoda and elevated them to a class which he named Monogenoidea
and combined with classes of cestodes to form superclass Cercomer-
morphae. Based largely on larval hooks, that concept embodies much
of the Cercomer Theory of Janicki (1920), who postulated homology of
the cercarial tail in the Digenea and the monogenean opisthohaptor
with the cercomer of larval cestodes. In contrast to that scheme,
Stunkard (1963) retained the monogeneans in the Trematoda but expressed their separation from other flukes by recognizing two subclasses
for which names proposed by Burmeister (1856) were used: Pectoboth-
ridia (equivalent to Monogenea of other schemes); and Malacobothridia
with two subclasses, Aspidobothrea and Digenea.

Intermediate to those schemes is one which accords to monogeneans the status of a class independent of the cestodes and Trematoda, and is implied by the suggestion that monogeneans descended from free-living ancestors in a line apart from that of the cestodes and digeneans. Because such descent remains a distinct possibility, this paper is limited to the aspidobothreans and digeneans. As used here, then, the terms Trematoda and trematodes arbitrarily exclude the monogeneans whose phylogeny and systematics have been reviewed most recently by Llewellyn (1965, 1970).

ORIGIN OF TREMATODES

Because trematodes are more evenly divided between marine and nonmarine species than any other class of parasitic worms, the phylogeny of marine forms cannot be considered apart from species with freshwater and terrestrial hosts. However, there seems to be general acceptance of the hypothesis that trematodes originated in the sea although Baer (1951) regarded the absence of their developmental stages in cephalopods and amphineurans as convincing evidence of their origin in fresh water. He cited the survival of marine cercariae in diluted seawater, reported by Stunkard and Shaw (1931), as evidence that "their natural euryhalinity has enabled them in the past to become adapted to seawater." Nevertheless, Stunkard and Shaw interpreted the ability of marine cercariae to survive and behave normally in dilutions as great as one-eighth seawater as evidence of migration from the sea to fresh water, and cited a study (Adolph 1925) which showed that marine organisms have a much greater tolerance for fresh water than freshwater organisms have for seawater.

Further indication of the marine origin of trematodes is provided by families with both marine and freshwater species, such as the Microphallidae, Lepocreadiidae, Opecoelidae, Microscaphiidae, and Pronocephalidae. Each family is represented by only a few freshwater species but has a number of marine representatives whose variety indicates a long association with marine hosts. That association probably began with prosobranch gastropods which were well represented and diversified in Cambrian marine fossils. Early Permian deposits give the oldest records of freshwater prosobranchs; pulmonates and freshwater lamellibranchs did not appear until the Cretaceous period.

After trematodes became established in freshwater as well as marine hosts, there may have been migration in both directions. A shift from marine to freshwater environments seems certain to have occurred more than one time in certain families. Thus, in the Haploporidae, a family otherwise restricted to marine hosts, species of Saccocoelioides have been found in freshwater fishes in Argentina (Szidat 1954), Texas (Lumsden 1963), and Australia (Martin 1973). In all three localities, the vertebrate hosts are not closely related to the fishes harboring most marine haploporids. Molluscan hosts of the Argentine species are unknown, but the close relationship of the snail serving the Texas species to marine prosobranchs led Cable and

Isseroff (1969) to conclude that the adaptation of marine molluscs to fresh water is a more decisive factor than the euryhalinity of vertebrate hosts in such trematodes transferring from a marine to a freshwater environment. They reported that in northwestern Indiana, several hundred miles from salt water, the hydrobiid snail *Amnicola limosa* is the host of a variety of cercariae, mostly of types more characteristic of marine than freshwater trematodes. Their occurrence far inland suggests a gradual change of habitat from marine to freshwater with the retreat of the seas that once inundated much of North America. However, the ability of marine snails to invade fresh water is suggested by the presence of a species of *Neritina*, a predominantly brackish water and marine genus, in mountain streams of Puerto Rico where the snails cannibalize one another's shells, presumably for scarce calcium (Ferguson 1959).

As to trematode ancestors, helminthologists have long regarded the saccate intestine of many monogeneans, all aspidobothreans, a few adult digeneans, and the rediae of many others as evidence of their descent from rhabdocoel turbellarians. Bresslau and Reisinger (1928) named the Graffilidae as the rhabdocoel family closest to ancestral monogeneans and the Anoplodiidae or a similar family for the digeneans. Recently, however, Wright (1971) has offered another hypothesis which suggests that the mesozoans, aspidobothreans, and digeneans evolved "from an unidentified, possibly protozoan ancestor" in a line independent of other flatworms. That hypothesis rests largely on the fact that all three groups parasitize molluscs, and on the assumption that the closest relatives of the trematodes are the mesozoans. While it is generally believed that the trematodes were first parasitic in molluscs, it is difficult to conceive, as Wright's hypothesis requires, that the many features they have in common with other flatworms are the result of parallel development or convergence.

TAXONOMY OF TREMATODES

Before many life histories were known, major taxa of the Digenea were based on the number and location of adhesive organs. The shortcomings of such systematics have been discussed at length elsewhere (Stunkard 1946) and need not be reviewed again here. Whether applied to adults or cercariae, basing higher categories on suckers has been thoroughly discredited. In a paper rarely cited, Ozaki (1937) regarded amphistomes as ancestral to other trematodes which he derived in three lines from the Paramphistomidae, Opistholebetidae, and Cephaloporidae. To the list of families with at least one amphistomate genus, our Caribbean studies (Nahhas and Cable 1964) added the Haplosplanchnidae, which has the further distinction of being one of the two families of digenetic trematodes uniformly having a rhabdocoel gut in the adult. Some amphistomes doubtless are more primitive than others and the same can be said of monostomes, but deriving other adult and larval trematodes from either type, as Ozaki (1937) and Sewell (1922) have done, leaves their polyphyletic nature unexplained.

Life history studies have profoundly altered concepts of families and higher categories of Digenea which were based solely on the morphology of adults by revealing that their convergence and divergence have sometimes masked actual relationships. A striking example of that situation is shown in Figure 1. The family Heterophyidae was erected to accommodate a number of small trematodes which were allocated to three subfamilies: Heterophyinae, Microphallinae, and Gymnophallinae, as represented within the circle. Distinction of the family and its subfamilies was based on such features as the genital modifications replacing the usual cirrus sac (shown at the bottom of Fig. 1). But when life cycles became known for all three subfamilies, it was discovered that each has a fundamentally different type of cercaria and life history. In the Heterophyinae, cercariae are produced by rediae in gastropods and nearly always penetrate a fish to encyst, whereas microphalline cercariae develop in sporocysts in gastropods and usually encyst in crustaceans. Moreover, gymnophalline larvae are produced by sporocysts in lamellibranchs and develop to unencysted metacercariae, usually in molluscs. These differences allied each subfamily to a different family and provided the basis for combining families with such diverse adults as those shown in Figure 1 to form the superfamilies Opisthorchioidea and Plagiorchioidea of the superorder Epitheliocystidia and the Brachylaemoidea of the superorder Anepitheliocystidia in the system proposed by La Rue (1957).

Basic to La Rue's system is the embryology of the cercarial stage. In some taxa, the tail is formed largely if not entirely by the molding of the embryo in such a manner that the primary excretory ducts are carried the length of the tail to open posteriorly or laterally on furcae, if present (as shown for gymnophalline larvae, in the lower segment of Figure 1). At the other extreme are cercariae in which the tail is formed entirely by proliferation of the region between the primary excretory pores so that they come to lie in the body-tail furrow (as shown for the microphalline cercaria, in the upper right segment of Figure 1). Other types of cercariae have the tail formed by a combination of molding and proliferation so that the excretory system extends into the tail for an appreciable distance to open at lateral pores (as shown for the heterophyine larva, in the upper left portion of Figure 1). Other features can be misleading, especially caudal ornamentation and rudimentation, but the manner in which the tail forms is determined by processes set in motion so early in ontogeny and so consistent for each type of cercaria that its significance for phylogenetic taxonomy seems beyond question. The conservative nature of that process is well demonstrated by the fellodistomids and brachylaemids whose larvae are basically furcocercariae. Even though some species lack the caudal furcae and others have the tail greatly reduced, the furcocercous stamp is evident from the caudal excretory tube as long as a vestige of the tail remains.

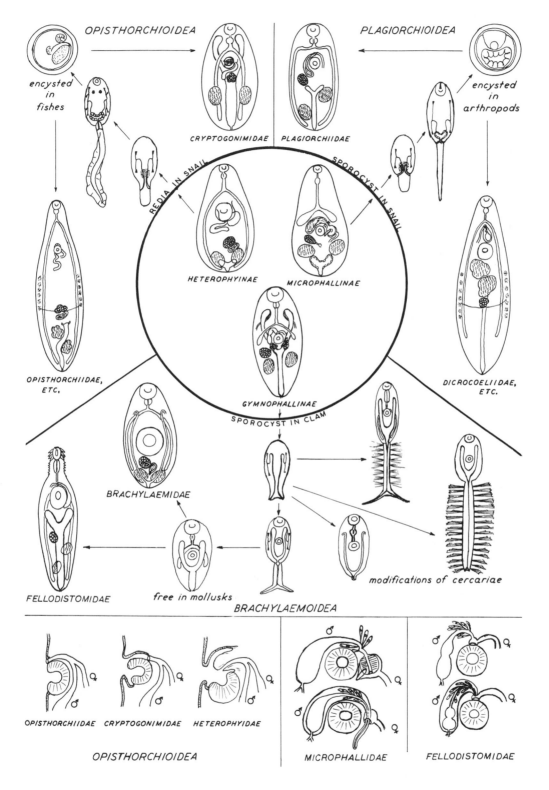

Figure 1

EVOLUTION OF DIGENETIC LIFE CYCLES

Most helminthologists with more than a casual interest in trematodes have indulged in speculation as to the origin and evolution of the complex life cycles in the Digenea. As cycles were demonstrated in group after group, taxonomic schemes were revised to incorporate the new information and evolutionary patterns which emerged. They have been interpreted in several publications (James and Bowers 1967; Ginetsinskaya 1968, 1971; Pearson 1972; and Rohde 1971b, 1972).

They agree that ancestors of the trematodes were first monoxenous parasites of archaic molluscs. Earlier, Baylis (1938) stressed the possibility that parasitism in molluscs alternated with a free-living adult stage. Heyneman (1960) postulated that a cercaria-like stage left the mollusc and encysted to overwinter before excysting and developing into the hermaphroditic adult. Cable (1965) suggested that the free-living adult was cercaria-like, with a natatory tail replacing the ancestral cilia used for locomotion. As actively swimming rather than creeping organisms, such primitive trematodes would have been well preadapted for occupying new niches including the bodies of larger animals and becoming dependent on them through the interplay of genetic and environmental factors. Rohde (1971b) considered the ready loss of the cercarial tail as evidence that the ancestral adult trematode was not caudate but, judging from the resemblance between redial and cercarial embryos of the bivesiculids, it seems likely that a sharp distinction between the body and tail was a later development.

The first step toward becoming heteroxenous parasites is believed to have occurred when fish ingested the free-living adult trematodes and they became facultative intestinal parasites. Host penetration would not have been involved and, at that stage, encystment would serve no purpose if, as seems likely, the heteroxenous cycle evolved in warm seas. But when the adult trematode became an obligate parasite no longer able to nourish itself adequately while free living, encystment in a benign environment would have tremendous survival value as is obvious from comparing the life spans of cercariae and metacercariae of trematodes that encyst in the open.

Penetration and encystment are not correlative. The ability to penetrate a second intermediate host seems likely to have begun as a superficial relationship between cercariae and animals eaten by the definitive host. A logical next step would have been for cercariae to be ingested by potential second intermediate hosts and to survive by leaving their intestines. This occurs today in a variety of helminths when they are swallowed at a stage other than the one normally continuing development in the intestine, or by a host that is unfavorable for such development. Evolution of the three-host cycle which is now characteristic of most digeneans was fully realized when their cercariae became able to penetrate more or less specific second intermediate hosts and to develop in this host into metacercariae infective for the vertebrate preying on those hosts. Encystment was not essential but usually occurred in the second intermediate host, and in a manner fundamentally different from that process in cercariae which encyst in the open on food of the vertebrate host or superficially in its prey. Moreover, it seems likely that encystment in specific second intermediate hosts evolved more than once to account for that process in such otherwise different groups of trematodes as the strigeoids and opisthorchioids, even though both groups encyst mostly in fish.

In some families, life cycles involve only two obligate host species and are clearly primitive. In several others, predominantly those with three-host cycles, some species that require only two hosts have secondarily simplified cycles in which precocious development in the molluscan host has eliminated the need for a second intermediate host and the free-living cercarial stage that infects it. Occasionally, however, the original vertebrate host has been eliminated by precocious development or progenesis in the second intermediate host, and very rarely, the entire cycle can be completed in the molluscan host species alone. In the opposite direction, a very few strigeoid trematodes have four-host cycles with an additional stage, the mesocercaria, developing in a host between the mollusc and the usual second intermediate host which may become a paratenic host (Pearson 1956).

PHYLOGENY OF THE DIGENEA

When information from all sources is brought to bear on the problem of phylogeny, it supports the hypothesis that the trematodes diverged early in their history and evolved independently in several lines. The writer's concept of what those lines may have been is diagramed in Figure 2. Only a select few of the nominal families of digeneans are included because of limited space and information. Some of those omitted are valid but many others are based on features of questionable family magnitude. Pearson (1972) did not propose such a scheme but his concepts of primitive life histories, their subsequent evolution, and the families cited as examples agree with the scheme presented here in many respects.

The number and diversity of digenean families that have basically furcocercous larvae and life histories ranging from the simplest to the most complex convince the writer that the furcocercaria was the primitive larval type. Of families having such larvae, the Bivesiculidae is believed to have retained more primitive features than any other family, and suggests that ancestral trematodes had free-living, furcocercous adults. Adults in that family are more like turbellarians than trematodes in three respects: (1) they lack adhesive organs of any sort; (2) the right and left excretory ducts may not fuse to form the single bladder characteristic of all other digeneans; and (3) the male copulatory organ is directed posteriorly and also is well removed from the end of the body. Bivesiculid

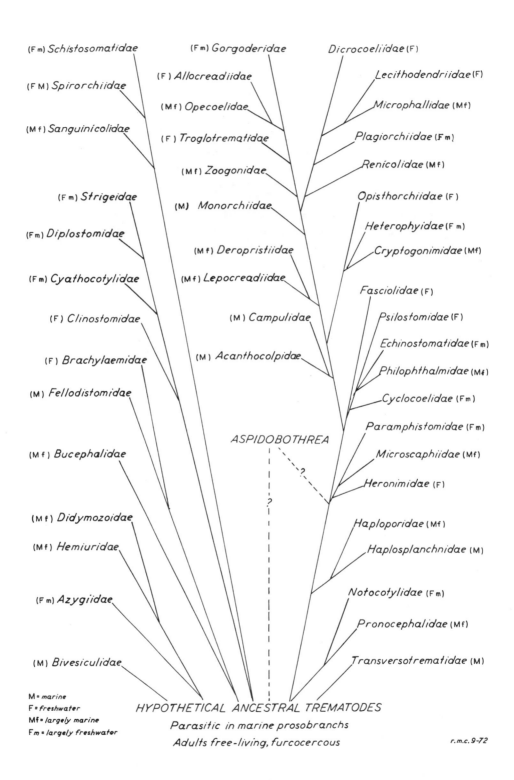

(Fm) *Schistosomatidae*

(FM) *Spirorchiidae*

(Mf) *Sanguinicolidae*

(Fm) *Strigeidae*

(Fm) *Diplostomidae*

(Fm) *Cyathocotylidae*

(F) *Clinostomidae*

(F) *Brachylaemidae*

(M) *Fellodistomidae*

(Mf) *Bucephalidae*

(Mf) *Didymozoidae*

(Mf) *Hemiuridae*

(Fm) *Azygiidae*

(M) *Bivesiculidae*

(Fm) *Gorgoderidae*

(F) *Allocreadiidae*

(Mf) *Opecoelidae*

(F) *Troglotrematidae*

(Mf) *Zoogonidae*

(M) *Monorchiidae*

(Mf) *Deropristiidae*

(Mf) *Lepocreadiidae*

(M) *Campulidae*

(M) *Acanthocolpidae*

Dicrocoeliidae (F)

Lecithodendriidae (F)

Microphallidae (Mf)

Plagiorchiidae (Fm)

Renicolidae (Mf)

Opisthorchiidae (F)

Heterophyidae (Fm)

Cryptogonimidae (Mf)

Fasciolidae (F)

Psilostomidae (F)

Echinostomatidae (Fm)

Philophthalmidae (Mf)

Cyclocoelidae (Fm)

Paramphistomidae (Fm)

Microscaphiidae (Mf)

Heronimidae (F)

Haploporidae (Mf)

Haplosplanchnidae (M)

Notocotylidae (Fm)

Pronocephalidae (Mf)

Transversotrematidae (M)

ASPIDOBOTHREA

?

?

M = *marine*
F = *freshwater*
Mf = *largely marine*
Fm = *largely freshwater*

HYPOTHETICAL ANCESTRAL TREMATODES
Parasitic in marine prosobranchs
Adults free-living, furcocercous

r.m.c. 9-72

Figure 2

cercariae are large, furcocystocercous larvae whose swimming attracts the attention of marine fish which swallow them and become infected. Instead of the extreme polymorphism of generations in the life cycles of other trematodes, bivesiculids have vurcocercous rediae with excretory pores at the tips of the furcae as in cercariae and, when young, are so similar to cercarial embryos that the two are easily confused when the infected snail is cracked. Finally, the specificity of bivesiculids for both their molluscan and vertebrate hosts seems unequaled by any other family. All of several known species of bivesiculid cercariae develop in snails of the genus *Cerithium*. Specificity for the vertebrate host may be less rigid but we found adults of each of two Caribbean species to have only one host species although the frequent occurrence of cercarial bodies in the intestine of various fish proved that they had been feeding on bivesiculid cercariae.

The family most like the Bivesiculidae in the general course of the life cycle is the Azygiidae. Larvae in that family are the largest of all cercariae and, like bivesiculid larvae, are furcocystocercous, have no penetration or cystogenous glands, and are swallowed by their vertebrate hosts which are fish. However, all azygiids are freshwater trematodes, except for species of *Otodistomum* which are among the few digeneans parasitizing elasmobranchs. Unlike bivesiculids, adult azygiids are distomes with powerful suckers and an excretory commissure dorsal to the oral sucker. Moreover, azygiid miracidia have spines instead of the usual cilia. In these respects, azygiids are most like hemiurids (Odhner 1911), and although their cercariae and life cycles differ considerably, there are similarities. The cystophorous larva of hemiurids has a tail that is much more complex but is modified so that the body is withdrawn into it, contains flame cells, and has an excretory appendage which bears the embryonic pores and resembles the tail of a furcocercaria in a species described by Chubrik (1952). Hemiurid cercariae also lack penetration and cystogenous glands and are eaten by the next host which, however, is not the vertebrate but a crustacean second intermediate host. In its intestine, a unique structure, the delivery tube, is forcibly ejected and the body of the cercaria passes through it in a manner suggesting that the tube may pierce the intestine of the crustacean and inoculate the cercarial body into the hemocoel. There the worm remains unencysted and develops to a metacercaria with well advanced features of the adult. Withdrawal of the body into the tail is a protective adaptation that seems to have evolved independently in the Bivesiculidae and azygiid-hemiuroid group as well as in the Gorgoderidae, all with cercariae that are eaten by the next host.

As to affinities of the Didymozoidae, there has been much confusion, but that group undoubtedly is digenetic and evolved from hemiuroid stock (Odhner 1907). A didymozoid life cycle has not yet been demonstrated experimentally but stages have been described that could scarcely be other than didymozoid cercariae developing in

marine snails, unencysted metacercariae in a crustacean second intermediate host, and juveniles in the intestine of fish that probably
serve as paratenic hosts (Cable and Nahhas 1962).

The Transversotrematidae, which may be the smallest family of
marine trematodes, also has a two-host cycle with furcocercariae
that have neither penetration nor cystogenous glands. However, the
body is not retracted into the tail and the cercaria is not eaten by
the vertebrate host as in bivesiculid and azygiid larvae. Instead,
the cercaria attaches to a suitable fish and the body develops to a
mature adult beneath the scales. The small size of the family, its
restriction to the Indo-Pacific region, and features of the life
cycle indicate an isolated position of the Transversotrematidae in
the phylogeny of the Digenea.

As to the Bucephalidae, that family was once separated taxonomically from all other digeneans but life history studies clearly
demonstrate its position to be with those having furcocercariae and
probably closest to the Fellodistomidae and Brachylaemidae. All
three families have small eggs and miracidia with cilia confined to
patches or appendages. And unlike all other trematodes with furcocercariae, the tail-stem lacks flame cells even though it is especially well developed in some fellodistomid larvae. In that group,
moreover, one known species has a tail remarkably like that of the
"ox-head" bucephalid larvae. Unique features of adult bucephalids
are the ventral position of the mouth and having a rhynchus instead
of an oral sucker at the anterior end. Also, the excretory system
is mesostomate whereas it is stenostomate in fellodistomids and
brachylaemids. The fact that the rhabdocoel intestine in bucephalids extends from the mouth and pharynx more toward the anterior end
of the body than posteriorly as in rediae and rhabdocoel turbellarians suggests that during evolution of the bucephalids, the mouth
shifted from near the anterior end of the body well back on the ventral surface. As a result, the primitive stenostomate form of the
excretory system became mesostomate and the relative position of
other structures was altered. Of the small number of trematode
families whose molluscan hosts are limited to lamellibranchs, the
Fellodistomidae is entirely marine and the Bucephalidae is largely
so.

In what seems to have been a separate line of descent from primitive trematodes are several families which have less modified furcocercariae with flame cells in the tail-stem and well-developed penetration glands. Also, their miracidia have two pairs of flame cells
instead of the usual single pair. That line diverged into two groups:
(1) the blood flukes with two-host cycles in which the cercaria penetrates the definitive host; and (2) another line including such families as the Clinostomidae and Strigeidae with cercariae which penetrate second intermediate hosts, nearly always fish, and usually
become encysted as metacercariae.

It has been theorized that the life cycle of blood flukes evolved from three-host cycles by development to adults in the second
intermediate host and elimination of the former definitive host. The

most abundant marine blood flukes are the sanguinicolids in fish.
Their cercariae develop in lamellibranchs or very rarely in annelids
whereas gastropods are the intermediate hosts of freshwater sanguinicolids. The Spirorchiidae is well represented by blood flukes of
marine turtles but known spirorchiid life cycles are limited to freshwater species which have cercariae developing only in pulmonate
snails. The schistosomatid blood flukes of birds and mammals are
predominantly freshwater species and develop in both prosobranchs
and pulmonates.

   The remaining trematodes are believed to have evolved from furcocercous ancestors in a line beginning with two changes: (1) loss
of the caudal furcae; and (2) acquisition of the ability to encyst
in the open. However, the tail continued to develop largely by
molding in the families of monostomes and amphistomes postulated at
one time or another to have been ancestral to other trematodes. With
evolution in that line, the cercarial tail formed more and more by
proliferation as indicated by the shortened caudal excretory tubule
in such families as Echinostomidae and Fasciolidae. Only in the
Heronimidae does the cercaria have flame cells in the tail, and
although Prevot (1971) found the larva of an echinostome to have a
small forked tail, it appears to be specialized for adhesion rather
than a reduced form of the primitive tail because the primary excretory pores seem to be on the stem rather than the furcae. It is
believed that the same type of tail occurs in a more reduced form in
the cyclocoelid described by Timon-David (1955). That species further resembles Prevot's echinostome by having a fully formed redia
in the miracidium as occurs also in philophthalmids, a family close
to the Echinostomidae.

   A characteristic common to families in this line except Heronimidae is the rapid encysting of cercariae (1) in the open on food,
(2) superficially in a second intermediate host, or (3) without leaving the molluscan host, and becoming metacercariae that require little or no further development to be infective for the vertebrate host.
Associated with encystment in most of these trematodes and no others
are the so-called bâtonnet cells containing rods shown by electron
microscopy to be scrolls of cyst material which are secreted by the
encysting worm and unroll to form a laminated inner cyst layer
(Dixon and Mercer 1964). In the Philophthalmidae, however, the inner
cyst is formed in that manner in species of *Parorchis* but not *Philoph-*
thalmus, and cystogenous glands of the bâtonnet type are not always
present in cercariae of other families in this line.

   A secondary line leading to the remaining digeneans is believed
to have diverged from the preceding one and to have been represented
first by the acanthocolpids, and then by the opisthorchioid families.
Common to those groups are biocellate cercariae with the excretory
system extending a distance into the tail, and conspicuously developed penetration glands. The larvae actively penetrate fish or rarely amphibians, encyst, and develop appreciably before they are infective for the vertebrate host. Some echinostome and philostome cercariae also encyst in fish, and their adults are very much like

acanthocolpids, even to the stenostomate form of the excretory system which seems to be retained as a primitive feature.

The Campulidae is comprised of liver and intestinal flukes of marine mammals which were once placed in the Fasciolidae and are still regarded by many helminthologists as being close to that family. A campulid life history has not yet been determined but from our unpublished observations on *Orthosplanchnus fraterculus* in the sea otter, it is clear that the cercaria is biocellate. The adult has many other features of the acanthocolpids which are obtained by eating infected fish, a major item in the diet of marine mammals that harbor campulids in contrast with the herbivorous hosts of fasciolids. Present information thus indicates that the Campulidae is closer to the Acanthocolpidae than any other family.

In all of the families in Figure 2 yet to be considered, cercariae have tails formed entirely or nearly so by proliferation. As a result, the excretory system is confined to the body and the primary pores open in the body-tail furrow. In all except the Lepocreadiidae, Deropristiidae, and Monorchiidae, cercariae have a stylet which is set vertically in the dorsal lip of the oral sucker in some families, and horizontally in others. The phylogenetic significance of stylets is uncertain. In the Acanthocolpidae and Renicolidae, a stylet is present in at least one species and absent in another apparently belonging to the same genus.

In the scheme proposed here, Monorchiidae is indicated as being exclusively marine and Zoogonidae predominantly so but with a few freshwater representatives. Heretofore, the view has been just the reverse with the freshwater genus *Asymphylodora* being assigned to Monorchiidae. However, morphology and host-parasite relationships strongly indicate that *Asymphylodora* is closer to the marine zoogonids, and with the freshwater genera *Triganodistomum* and *Lissorchis*, links the Zoogonidae with the plagiorchioid line.

For the most part, the scheme shown in Figure 2 groups families much as La Rue (1957) did in allocating them to higher categories in his system, but there are exceptions. One is the Renicolidae for which he erected a new order, suborder, and superfamily, all of which have been invalidated by life history studies revealing that the renicolids are to be considered a family in the superfamily Plagiorchioidea of La Rue's order Plagiorchiida. A second exception is to his allying the Cyclocoelidae to trematodes having furcocercariae, and to expressing that relationship by erecting the suborder Cyclocoelata of the order Strigeatoidea. That action was prompted by resemblance of the cercarial tail in *Pseudhyptiasmus dollfusi* to a reduced fork tail (Timon-David 1955), but, as discussed above, that cyclocoelid has more in common with echinostomes, especially *Aporchis massiliensis*, whose life cycle was reported by Prevot (1971). A third exception concerns the Campulidae, which La Rue assigned to the order Echinostomida but is placed nearer the Acanthocolpidae in Figure 2. Those families comprise the suborder Acanthocolpiata, which Nahhas and Cable (1964) erected in the order Opisthorchiida after Peters (1961) showed that the embryology of the excretory system in the cercaria excluded acanthocolpids from the order Plagiorchiida where La Rue had placed them.

Farthest removed from their location in La Rue's system by the
scheme in Figure 2 are the hemiurids and didymozoids, which he placed
in the suborder Hemiurata, order Opisthorchiida, but for reasons
given above are placed closer to trematodes with furcocercariae and
thus comprising his order Strigeatoidea. As revealed in discussions
with La Rue before his system was published, he suspected affinities
between the hemiurids and strigeatoids. However, he gave more weight
to evidence that the hemiurid bladder is cellular in using structure
of the bladder to classify digeneans into two superorders, the Anepitheliocystidia and the Epitheliocystidia. Subsequent studies have
raised doubt as to that separation or at least as to his allocation
of families to superorders based on cellular and noncellular bladders.
First, Freeman and Llewellyn (1958) reported the bladder to be epithelial in the progenetic metacercaria of a trematode in the Fellodistomidae, a family which La Rue assigned to the Anepitheliocystidia.
More recently, Krupa, Cousineau, and Bal (1969) studied the fine
structure of the bladder in the cercaria of *Cryptocotyle lingua*, a
species in the Heterophyidae, a family assigned to the Epitheliocystidia. They found no indication that the bladder wall was cellular;
instead, it had the appearance of a glandular syncytium in which protein synthesis was conspicuous. In some trematode families, the bladder is thin-walled and inconspicuous in the cercaria but becomes
greatly thickened in the metacercaria with inclusions which are void-
ed through the excretory pore as the worm excysts or soon thereafter.
In such metacercariae, it is clear that the bladder wall is distended
with material for which the worm has no further need, presumably
metabolic wastes that cannot pass the barrier of the cyst wall.
Apparently, there would be little necessity for that function in
developing cercariae or in metacercariae that do not encyst. Moreover, protein synthesis in the bladder wall of *C. lingua* cercariae
indicates a function beyond excretion alone. For the Heterophyidae
at least, Leong and Howell (1971) may have provided the answer in
reporting that the excretory bladder is a source of cyst material in
the cercaria of *Stictodora lari*. Extrusion of cyst material from the
excretory pore had previously been reported for the Gorgoderidae
(Goodchild 1943) and may well occur in other families in which cercariae have bladders with thick granular walls. A comparative study
of the fine structure of cercariae in several selected families should
do much to clear up the matter of cellular versus noncellular bladders and to determine whether the digeneans are separable on that
basis into only two groups as La Rue postulated or into more than
two, which would correspond more or less to the orders in his system.

THE ASPIDOBOTHREANS

Because of their direct development, the aspidobothreans are
monogenetic, but their morphology and host relationships reveal much
closer affinities with the Digenea than with the Monogenoidea. They
were, in fact, included in the Digenea until Faust and Tang (1936)
proposed for them a separate subclass, the Aspidogastrea, coordinate

with the Digenea and Monogenea. Closer affinity of the Aspidobothrea and Digenea was later expressed in the schemes mentioned at the
beginning of this paper. Stunkard (1963) defined the aspidobothreans as "an aberrant, isolated group of parasites of mollusks and
of lower vertebrates which feed on such mollusks and infect both
marine and freshwater hosts in all parts of the world." Stunkard
continued, "It appears that the aspidogastrids and digenetic forms
have descended from a common turbellarian-like ancestor which initially became parasitic in mollusks and that the aspidogastrids never
acquired axexual methods of reproduction." The widespread occurrence
of progenesis in digeneans was cited as support of the thesis that
the direct life cycle of aspidobothreans is primitive and not derived
by loss of the "asexual" generations characteristic of digeneans.
That view does not exclude the possibility that present-day aspidobothreans are in part heteroxenous and in part progenetic descendents
of heteroxenous ancestors. Usually, however, they are held to be
primitively monoxenous, as restated most recently by Rohde (1971a,
b) who considered the group as "primitive, direct descendents of
turbellarians which are not yet closely adapted to parasitism and
have not yet incorporated the vertebrate host as a fixed component
of their life-cycle." In support of that interpretation, Rohde
cited the attainment of maturity in the molluscan host by some
aspidobothreans, notably *Aspidogaster conchicola*, which Huehner and
Etges (1972) have demonstrated to be transmissible from mollusc to
mollusc by ingestion of embryonated eggs. Rohde used eggs and larvae of *Multicotyle purvisi* from adults in freshwater turtles to
infect snails in four families. He demonstrated that only juvenile
worms that had developed to a considerable degree in snails were
infective for turtles, but he did not prove that the species can
develop to maturity and produce viable eggs in molluscs as required
by his interpretation of the live cycle.

Most aspidobothrean species are now known to occur naturally in
both molluscs and vertebrates. The most recent addition to that
list is *Lophotaspis interiora*, previously known only from turtles
but found by Hendrix and Short (1972) to occur only as juveniles in
twelve species of freshwater unionids in Florida. The genus *Lophotaspis* is of particular interest here because its six species are
widely distributed and evenly divided between marine and freshwater
hosts. The only species reported to complete its life history in a
mollusc parasitizes a freshwater clam in Japan.

Another aspidobothrean genus with both marine and freshwater
species is *Cotylogaster*, a genus including two species from marine
fish and one freshwater representative, *C. occidentalis*, which was
first reported from a fish, the sheepshead. Later, Dickerman (1948)
found in "*Goniobasis* snails" specimens with which he was able to
infect the sheepshead but not the rock bass. He reported that large
worms from snails contained many viable eggs, making a vertebrate
host unnecessary. Wooton (1966) obtained adults from the intestine
of freshwater lamellibranchs, hatched larvae from eggs teased from
the worms, and infected clams experimentally. We find *C. occidentalis* occurring singly as large individuals in the digestive gland of

about 2% of *Pleurocera acuta* in the upper Wabash watershed. The worm nearly always occupies the apical coil of the gland and replaces its tissues so that the bright orange color of the parasite seen through the tunic of the gland reveals the infection at a glance when the shell is cracked. Eggs are rarely present and then in small numbers so that the extensive uterus is largely empty. Moreover, we have never observed eggs or larvae of *C. occidentalis* in the water in which many thousands of *P. acuta* have been isolated to detect trematode infections at all seasons over a period of more than thirty years. Further indication that the life cycle is not normally completed in *P. acuta* at least is the genetic cul-de-sac imposed on the worm by its occurring singly in that host with no opportunity of cross fertilization. Elsewhere in that snail, we find *Cotylaspis cokeri*, also a common intestinal parasite of the turtle, *Graptemys geographica*, which feeds largely on molluscs but has not yet been found to harbor adults of *Cotylogaster occidentalis* .

The two most likely possibilities as to the origin of the aspidobothreans are indicated in Figure 2. One is that they and the digeneans diverged from a common ancestral stock which was monoxenous, monogenetic, and parasitic in archaic molluscs. The other is that they evolved later from digenetic and very likely heteroxenous ancestors. A third possibility, separate descent from a common turbellaria stock, seems less likely because of the features common to both digenetic and aspidobothrean trematodes and stressed most recently by Rohde (1972). One seemingly fundamental difference is the type of larva that hatches from the egg and infects the molluscan host. In the Digenea, it is the well-known miracidium which in some taxa differs considerably from the usual larva clad with tiers of ciliated cells. Because the larval aspidobothrean resembles a miniature amphistome with the structure of a redia more than of a miracidium, Wooton (1966) proposed the name cotylocidium for that stage. Cilia usually are present but are restricted to a definite pattern of small scattered areas probably corresponding to the so-called epidermal cells or plates of miracidia. And whereas the miracidium functions as a germinal sac, the cotylocidium develops directly to the hermaphroditic adult aspidobothrean.

Differences between the miracidium and the cotylocidium may be less significant than they seem. The structural gap between them is narrowed greatly by the observations of Ogambo-Ongoma (1970), reported in a thesis made available by Dr. John D. Goodman who directed the study. Laboratory reared snails of the normal host species of *Fasciola gigantica* in Uganda were exposed to miracidia and examined daily thereafter. After penetrating the snails, miracidia did not become sporocysts as expected. Instead, they developed by the fourth day from the miracidial stage to rediae typical of the usual mother redia in every respect.This study, made in a manner that evidently left no chance of error in results or their interpretation, suggests reexamination of the largely discarded view that the apical gland of the miracidium is a rudimentary intestine, and adds significantly to the evidence that all generations in the digenean cycle, however different they may appear, are fundamentally alike, as Looss (1892) clearly perceived and Cable (1965) emphasized.

The widespread occurrence of precocious development in the digeneans, the plasticity of phenotypic expression as evidenced by the
polymorphism of generations, and the known instances in which the
usual reproductive function of one generation is performed by another
make evolution of the cotylocidium from the miracidium, and descent
of the aspidobothreans from digeneans by no means as farfetched as
they may seem. Should the miracidium of *Fasciola gigantica* have
developed before hatching from the egg to the point that it does in
only four days after penetrating the snail, it would resemble a cotylocidium except for germinal development and lacking the posterior
sucker (which enables the cotylocidium to adhere to host membranes
and develop without penetrating them as miracidia do). Indeed, the
relatively superficial location of most aspidobothreans in their
molluscan hosts seems to be the only explanation of their remarkably
low degree of specificity for their mollsucan hosts and sites within
those hosts as compared with digeneans.

The digenean with the simplest known life cycle and thus most
like the aspidobothrean cycle is the freshwater turtle lung-fluke,
Heronimus mollis. Crandall (1960) found that the miracidium penetrated a pulmonate snail and developed rapidly in the mantle region
to a large sporocyst which produced a relatively small number of
amphistome cercariae with well-developed tails. They escaped from
the sporocyst but remained unchanged in the snail to be eaten by the
turtle. Freed when the snail is crushed in the turtle's mouth, the
cercariae migrate through the glottis to the trachea and lungs. The
adult lays no eggs but retains them until the next summer when the
gravid worm apparently migrates to the throat of the turtle and
escapes. The miracidium is large, advanced in germinal development,
and contains a number of cercarial embryos when it hatches from the
egg. Such "overripeness" may be associated with the absence of the
two or three redial generations that are characteristic of paramphi-
stomes, the trematodes closest to *H. mollis* and also often suggested
as being the nearest relatives of the aspidobothreans (Cable 1965).
Whether the simplicity of the heronimid cycle with only two generations is primitive or secondary is a moot question. That cycle differs in two respects from what seems to be secondarily simplified
ones in other families: (1) the cercaria does not encyst in the
snail host; and (2) the tail is not reduced but is an efficient swimming organ when cercariae are released by cracking the snail. But
whether the heronimid cycle is primitive or not, it is just one "step"
from the cycles of the aspidobothreans so far as generations are concerned and equally close in its host-parasite relationships.

Comparative morphology of the nervous system, including fine
structure of photoreceptors, may provide further clues to phylogeny
of the aspidobothreans. In digeneans, miracidia are more conservative than cercariae in structure of eyespots and usually show bilateral asymmetry, which in *H. mollis* extends to the presence of two
pigment cells on the left and one on the right. Other miracidia
investigated thus far have only one cell on each side (Pond and Cable
1966; Isseroff and Cable 1968). Asymmetry to that extent also occurs

in the cotylocidium of *Cotylogaster basiri*, a marine aspidobothrean
in which, however, fine structure of the photoreceptors is unknown.
Rohde (1971a, b) did not include that aspect in reporting detailed
studies of *Multicotyle purvisi*, a freshwater species, but has stated
in a personal communication that he did not recall asymmetry of
photoreceptors in the cotylocidium.

FURTHER STUDIES

Much remains to be learned from additional studies along the
lines of those responsible for current concepts of trematode taxonomy
and evolution. Despite their impact on those concepts, life histories
are still unknown for a number of families, and a better grasp of
comparative morphology would help clarify the systematic position of
many species and distinguish between unspecialized and derived features. And barely a beginning has been made in utilizing electron
microscopy.

A promising field of investigation apparently not yet applied to
trematodes is comparison of the amount and complexity of DNA in different species. The rationale is that unspecialized organisms have
more DNA than specialized ones which have lost abilities no longer
needed but essential for less specialized forms. Confirmation of
that thesis has been reported for teleost fish by Hinegardner and
Rosen (1972), who have postulated evolutionary lines on the basis of
nuclear DNA content as determined analytically. Using DNA hybridization, Searcy (1970) determined that the genome was simpler in parasitic than nonparasitic plants. That observation was confirmed by
Searcy and MacInnis (1970), using renaturation of denatured DNA; the
rate is inversely related to genome complexity. With that method,
however, they found DNA to be twice as complex in a cestode as in a
planarian, and distinctly more complex in two parasitic nematode
species than in two that are free-living. It was speculated that in
the evolution of parasites, there is at first a genetic reduction
which is followed by an accrual of new genetic information in adaptation to increased complexity of the host-parasite relationship. It
seems safe to predict that such methods and others yet to be discovered will eventually make precise genome comparisons possible, thereby indicating with much more certainty than present information off-
ers, the origin and subsequent evolutionary lines of the digenetic
trematodes.

SUMMARY

The digeneans and aspidobothreans are believed to have evolved
from a common ancestral stock of marine rhabdocoels that became parasitic as juveniles in primitive molluscs, probably prosobranch gastropods. Adults were free-living, pelagic forms with the posterior

body region bifurcated, modified for swimming, and eventually becoming the cercarial tail. They were eaten by fishes and became facultative intestinal parasites. When they were no longer able to nourish themselves adequately in the open and develop there to sexual maturity, parasitism became obligatory. Meanwhile, environment in the mollsucan host favored the development of juveniles to germinal sacs which produced cercariae. The life cycle thus became essentially as it is today in the Bivesiculidae, a marine family with more primitive features than any other trematodes. The life cycle is similar in the Azygiidae which are in another line leading also to the Hemiuridae and Didymozoidae (whose cercariae are eaten by a second intermediate host instead of by the fish that serve as definitive hosts). The Transversotrematidae is an isolated group whose cercar-
    Other furcocercariae developed penetration glands and evolved in two lines. One included the Fellodistomidae and Brachylaemidae (with unencysted metacercariae in invertebrates) and apparently the Bucephalidae (whose cercariae penetrate and encyst in fish). The other led to the clinostomid and strigoid families (with metacercariae usually in fish) and the families of blood flukes (which probably develop to maturity in what were once vertebrates serving as second intermediate hosts). The remaining trematodes are believed to form a single line of descent that began with cercariae losing the caudal furcae and acquiring cystogenous glands for encystment in the open, superficially on food, or in a second intermediate host (as occurs in the predominantly or exclusively marine families Haplosplanchnidae, Haploporidae, Pronocephalidae, and probably Microscaphiidae, and in mainly the freshwater digeneans in the Notocotylidae, Paramphistomidae, and the families of the echinostomoid complex, including the Cyclocoelidae and culminating in the Fasciolidae). A secondary line diverged from that complex to give rise first to the Acanthocolpidae and the Campulidae, then the opisthorchioid families, and finally those with cercariae bearing stylets. Whether the aspidobothreans diverged from the digeneans before or after the life cycle became heteroxenous and digenetic is uncertain. The simplicity of the aspidobothrean cycle, usually interpreted as being primitive, is approached in the Digenea by the cycle in the Heronimidae. Similarities in morphology and host-parasite relationships give credence to the possibility that the aspidobothreans evolved from digeneans as a result of progenesis and simplification of the life cycle.

LITERATURE CITED

Adolph, E. F. 1925. Some physiological distinctions between freshwater and marine organisms. Biol. Bull. 48:327-35.
Baer, J. G. 1951. Ecology of animal parasites. Urbana: Univ. of Illinois Press.
Baylis, H. A. 1938. Helminths and evolution. In Evolution; Essays on aspects of evolutionary biology, ed. G. R. De Beer, pp.249-70.

Bresslau, E., and Reisinger, E. 1928. Allgemeine Einleitung zur Naturgeschichte der Platyhelminthes. In Handbuch der Zoologie, ed. W. Kükenthal and T. Krumbach, 2:34-51.

Burmeister, H. 1856. Zoonomische Briefe. Allgemeine Darstellung der Thierischen Organization 1:1-367. Leipzig.

Bychowsky, B. E. 1937. Ontogeny and phylogeny of parasitic platyhelminths (in Russian). Izv. Akad. Nauk SSSR, Sci. Math. Nat. 1353-83.

_____. 1957. Monogenetic trematodes, their systematics and phylogeny. Akad. Nauk SSSR (in Russian). English transl. 1961, V.I.M.S. Trans. Ser. No. 1. Washington D. C.: A.I.B.S.

Cable, R. M. 1965. "Thereby hangs a tail." J. Parasitol. 51:3-12.

_____, and Isseroff, H. 1969. A protandrous haploporid cercaria, probably the larva of Saccocoelioides sogandaresi Lumsden, 1963. Proc. Helm. Soc. Wash. 36:131-35.

_____, and Nahhas, F. M. 1962. Lepas sp., second intermediate host of a didymozoid trematode. J. Parasitol. 48:34.

Chubrick, G. K. 1952. Cystophorous cercariae from Natica clausa Brod. and Sow. (in Russian). Dokl. Akad. Nauk SSSR 86:1233-36.

Crandall, R. B. 1960. The life history and affinities of the turtle lung fluke, Heronimus chelydrae MacCallum, 1902. J. Parasitol. 46:289-307.

Dickerman, E. E. 1948. On the life cycle and systematic position of the aspidogastrid trematode, Cotylogaster occidentalis Nickerson, 1902. J. Parasitol. 34:164.

Dixon, K. E., and Mercer, E. H. 1964. The fine structure of the cyst wall of the metacercaria of Fasciola hepatica. Quart. J. Mic. Sci. 105:385-89.

Faust, E. C., and Tang, C. C. 1936. Notes on new aspidogastrid species, with a consideration of phylogeny of the group. Parasit. 28:487-501.

Ferguson, F. F. 1959. Intraspecific predation in a Puerto Rican neretid snail. Trans. Am. Microscop. Soc. 78:240-42.

Freeman, R. F., and Llewellyn, J. 1958. An adult digenetic trematode from an invertebrate host: Proctoeces subtenuis from the lamellibranch Scrobicularia plana (Da Costa). J. Mar. Biol. Ass. U. K. 37:435-57.

Ginetsinskaya, T. A. 1968. Trematodes and their life cycle, biology and evolution (in Russian). Akad. Nauk SSSR, Leningrad.

_____. 1971. Theoretische Fragen des Lebenszyklus der Verwandschaftsbeziehungen und der Polyogenese der Trematoden. Parasit. Schriftenr. 21:11-16.

Goodchild, C. G. 1943. The life-history of Phyllodistomum solidum Rankin, 1937, with observations on the morphology, development and taxonomy of the Gorgoderinae (Trematoda). Biol. Bull. 84: 59-86.

Hendrix, S. S., and Short, R. B. 1972. The juvenile of Lophotaspis interiora Ward and Hopkins, 1931 (Trematoda: Aspidobothria). J. Parasitol. 58:63-67.

Heyneman, D. 1960. On the origin of complex life cycles in the digenetic flukes. Libro Hom. Caballero, pp. 133-52.

Hinegardner, R., and Rosen, D. E. 1972. Cellular DNA content and the evolution of teleostean fishes. Am. Nat. 106:621-44.

Huehner, M. K., and Etges, F. J. 1972. Experimental transmission of Aspidogaster conchicola von Baer, 1827. J. Parasitol. 58: 109.

Isseroff, H., and Cable, R. M. 1968. Fine structure of photoreceptors in larval trematodes. Z. Zellf. 86:511-34.

James, B. M., and Bowers, E. A. 1967. Reproduction in the daughter sporocyst of Cercaria Bucephalopsis haimeana (Lacaze-Duthiers, 1854) (Bucephalidae) and Cercaria dichotoma Lebour, 1911 (non Müller) (Gymnophallinae). Parasitol. 57:607-25.

Janicki, C. 1920. Grundlinien einer "Cercomer" Theorie zur Morphologie der Trematoden und Cestoden. Festschr. Zschokke No. 30.

Krupa, P. L., Cousineau, G. H., and Bal, A. K. 1969. Electron microscopy of the excretory vesicle of a trematode cercaria. J. Parasitol. 55:985-92.

La Rue, G. R. 1957. The classification of digenetic trematodes: A review and a new system. Exp. Parasitol. 6:306-44.

Leong, C. H. D., and Howell, M. J. 1971. Formation and structure of the cyst wall in Stictodora lari (Trematoda: Heterophyidae). Z. Parasitenk. 35:340-50.

Llewellyn, J. 1965. The evolution of parasitic platyhelminths. In Evolution of parasites, ed. A. E. R. Taylor, pp. 47-78. Third Symp. Brit. Soc. Parasitol. Oxford: Blackwell Sci. Pub.

_____. 1970. Monogenea. 2nd Intl. Cong. Parasitol. J. Parasitol. 56(4, Sec. II, Part 3):493-504.

Looss, A. 1892. Ueber Amphistomum subclavatum Rud. und seine Entwicklung. In Festschr. 70 Geb. R. Leuckarts, pp. 147-67. Leipsig.

Lumsden, R. D. 1963. Saccacoelioides sogandaresi sp. n., a new haploporid trematode from the sailfin molly, Mollienisia latipinna Le Sueur in Texas. J. Parasitol. 49:281-84.

Martin, W. E. 1973. Life history of Saccocoelioides pearsoni n. sp. Trans. Am. Microscop. Soc. 92:80-95.

Nahhas, F. M., and Cable, R. M. 1964. Digenetic and aspidogastrid trematodes from marine fishes of Curacao and Jamaica. Tulane Stud. Zool. 11:168-228.

Odhner, T. 1907. Zur Anatomie der Didymozoen: ein getrenntgeschlechtlicher Trematode mit rudimentärem Hermaphroditidmus. Zool. Stud. Tillägn. T. Tullberg, pp. 309-42.

_____. 1911. Zum natürlichen System der digenen Trematoden 4. Zool. Anz. 38:513-31.

Ogambo-Ongoma, A. H. 1970. Some aspects of fascioliasis in Uganda. Thesis, Fac. Agric. Kampala: Makere Univ.

Ozaki, Y. 1937. Studies on the trematode families Gyliauchenidae and Opistholebetidae with special reference to lymph system. J. Sci. Hiroshima Univ. Ser. B., Div. 1, 5:167-242.

Pearson, J. C. 1956. Studies on the life cycles and morphology of the larval stages of *Alaria arisaemoides* Augustine and Uribe, 1927, and *Alaria canis* La Rue and Fallis, 1938 (Trematoda: Diplostomidae). Canad. J. Zool. 34:295-387.

_____. 1972. A phylogeny of life-cycle patterns of the Digenea. Adv. Parasitol. 10:153-89.

Peters, L. E. 1961. The allocreadioid problem with reference to the excretory system in four types of cercariae. Proc. Helm. Soc. Wash. 28:102-108.

Pond, G. G., and Cable, R. M. 1966. Fine structure of photoreceptors in three types of ocellate cercariae. J. Parasitol. 52: 483-93.

Prevot, G. 1971. Cycle évolutif d'*Aporchis massiliensis* Timon-David, 1955 (Digenea: Echinostomatidae), parasite du goeland *Larus argentatus michaellis* Naumann. Bull. Soc. Zool. France 96: 197-208.

Rohde, K. 1971a. Untersuchungen an *Multicotyle purvisi* Dawes, 1941 (Trematoda: Aspidogastrea). I. Entwicklung und Morphologie. Zool. Jahrb. Anat. 88:138-87.

_____. 1971b. Phylogenetic origin of trematodes. In Perspektiven der Cercarienforschung, ed. K. Odening, Parasitol. Schriftenr 21:17-27.

_____. 1972. The Aspidogastrea, especially *Multicotyle purvisi* Dawes, 1941. Adv. Parasitol. 10:77-151.

Searcy, D. G. 1970. Measurements by DNA hybridization in vitro of the genetic basis of parasitic reduction. Evolution 24:207-19.

_____, and MacInnis, A. J. 1970. Measurements by DNA renaturation of the genetic basis of parasite reduction. Evolution 24:796-806.

Sewell, R. B. S. 1922. Cercariae Indicae. Indian J. Med. Res 10 (Suppl.) pp. 1-370.

Stunkard, H. W. 1963. Systematics, taxonomy and nomenclature of the trematoda. Quart. Rev. Biol. 38:221-33.

_____, and Shaw, C. R. 1931. The effect of dilution of sea water on the activity and longevity of certain marine cercariae, with descriptions of two new species. Biol. Bull. 61:242-71.

Szidat, L. 1954. Trematodos neuvos de peces de agua dulce de la Republica Argentina y un intento para aclarar su character marino. Rev. Inst. Nac. Invest. 3:1-82.

Timon-David, J. 1955. Cycle évolutif d'un trématode cyclocoelide: *Pseudhyptiasmus dollfusi* Timon-David, 1950. Ann. Parasitol. 30:43-61.

Wooton, D. M. 1966. The cotylocidium larva of *Cotylogasteroides occidentalis* (Nickerson, 1902) Yamaguti, 1963 (Aspidogasteridae-Aspidocotylea-Trematoda). Proc 1st Intern. Congr. Parasitol. 1:547-58.

Wright, C. A. 1971. Flukes and snails. Sci. of Biol. Ser., No. 4. London: Geo. Allen and Unwin.

Gyrocotyle: a century-old enigma

J. E. Simmons
Department of Zoology
University of California
Berkeley, California

The genus *Gyrocotyle* was established by Diesing in 1850, and questions relating to this interesting group of flatworms were present at the outset, for Diesing's study material was erroneously thought to have been obtained from *Antilope pygarga*, a South African ungulate. Gyrocotylideans have subsequently been taken only from chimaeroid fish (except for the very rare adult specimen recovered free from the host) of which they are characteristic helminth parasites.

Much of the earlier work with gyrocotylideans was devoted to their morphology and taxonomic status, and to the question of their antero-posterior orientation. Excellent historical accounts and summaries of the older studies are available (Watson 1911; Fuhrmann 1930). The basic morphology of gyrocotylideans has been thoroughly described and the problem of orientation has long since been settled to the satisfaction of most helminthologists. However, many details of the biology of the group, particularly (1) that portion of the life history from larva to adult, (2) whether an intermediate host is utilized, (3) metamorphosis and growth, and (4) governance of infection intensities, remain unknown and are of considerable intrinsic interest. Also, the systematic postion and affinities of gyrocotylideans may still be said to be controversial. Some of the more unusual or enigmatic features of gyrocotylidean biology will be briefly described and discussed in this paper.

SPECIATION

As far as is known, each species of *Gyrocotyle* is restricted to a single species of chimaeroid host (van der Land and Dienske 1968). Thus, although numerous authorities (Watson 1911; Lynch 1945) here indicated that a species of *Gyrocotyle* from *Hydrolagus colliei* of the Pacific coast of the United States is perhaps identical to *Gyrocotyle urna* from *Chimaera monstrosa* of European waters, van der Land and Dienske (1968) have provided evidence that the former is distinctive and have redescribed it as *Gyrocotyle parvispinosa* Lynch, 1945. (This is perhaps an unfortunate choice of specific name, and the two varieties noted by Lynch, a very careful observer, forma *parvispinosa* and forma *magnispinosa*, deserve additional investigation.)

Further, van der Land and Dienske (1968) have pointed out two additional features of *Gyrocotyle* parasitism which are most interesting. These authors maintain that in each host species a second, and rarer, species of *Gyrocotyle* occurs, in addition to the common one. Thus, *Chimaera monstrosa* may be infected with *G. urna* or *G. confusa* (a related gyrocotylidean, *Gyrocotyloides nybelini* Fuhrmann, 1931 also occurs in *C.monstrosa*), *Hydrolagus colliei* with *G. fimbriata* or *G. parvispinosa*, *Hydrolagus affinis* with *G. major* or *G. abyssicola* (van der Land and Templeman 1968), and so forth. This fact has probably been obscured for many years because no more than two dozen specimens of *G. confusa* have been recorded (Dienske 1968). We have pointed out elsewhere (Simmons and Laurie 1972) that the insistence of Ward (1912) and Wardle (1932) that *G. fimbriata* was not a valid species is untenable. This worm clearly differs from *G. parvispinosa* in morphology (Watson 1911; Lynch 1945; van der Land and Dienske 1968), in intestinal distribution (Simmons and Laurie 1972), in certain characteristics of its permeability to carbohydrates (Laurie 1971), and in degree of heteroduplexing of DNA (Simmons et al. 1972).

At present it is thought that the common *Gyrocotyle* from all the host species form one species complex, and the less common, or rare, species form another (van der Land and Dienske 1968). The anticipated systematic survey of the subclass Gyrocotylida (Dienske 1968) by van der Land will doubtless greatly clarify and elaborate upon the problems of speciation and host specificity exhibited by represenatives of this group of flatworms.

INFECTION INTENSITIES

Escape of *Gyrocotyle* from dead or moribund hosts has been recorded by a number of workers, and the rapidity with which this occurs has been documented (Simmons and Laurie 1972). Despite this, workers with even a modest number of host animals at their disposal have found that the infections most commonly encountered are those consisting of a pair of worms. Dienske (1968) has conveniently summarized the most important records of *G. urna* from *Chimaera monstrosa*, and we have established, in a careful examination of over 400 *Hydrolagus colliei* that 68% of the infections with *G. fimbriata* and 56% of those with *G. parvispinosa* were comprised of a pair of worms (Simmons and Laurie 1972).

Two questions obviously arise: Why is there so frequently a pair of worms? How do the worms become established and maintained as a pair? In answer to the first of these questions, Halvorsen and Williams (1968) cite several studies in which a similar phenomenon has been noted with other intestinal helminths and, of course, there are cases in which parasites are coupled by tissue fusion and become encysted as pairs. These workers suggest that the two *Gyrocotyle* comprise a functional sexual unit; this number is the minimal one which would insure cross fertilization. This suggestion is an excellent one. Gyrocotylideans exhibit protandry, and it is worth noting that the vaginal pore is located on the dorsal aspect of the worm, while the male pore is situated ventrally, and there is no permanent, protrusible intromittent organ (Lynch 1945).

The mechanism regulating the infection intensity of *Gyrocotyle* is debatable. It has frequently been suggested that the worm pair is selected from an initial infection with a larger number of young. So plausible indeed is this explanation that in compendia it sometimes appears as an incontrovertible fact; thus, Wardle and McLeod (1952, p. 659) state that "it occurs in pairs, representing twin survivors of a mass infection. . .".

Dienske (1968), however, submitted a collection of *G. urna* to analysis and found the infection intensities to be compatible with a Poisson distribution erected upon the hypothesis that worms enter the host one by one at random, but to the exclusion of any third parasite. Supernumerary infections do occur and indicate that the restrictive mechanism, whatever its nature, is imperfect in operation. In our own study of gyrocotylideans from *Hydrolagus colliei* (Simmons and Laurie 1972), we have distinguished between supernumerary infections and the so-called "massive" infections with large numbers of postlarval forms. The latter may consist of over 200 individuals (Lynch 1945). Infections of this nature are very rare and in our own experience have not been found in the youngest fish; rather, such hosts reveal a greatly enhanced incidence of single-worm infections. Evidence for a suppressive mechanism, as suggested by Dienske (1968), is further provided by the exceedingly rare occurrence of mixed infections of *G. fimbriata* and *G. parvispinosa*. This indicates that infections are not randomly acquired and that they appear to influence one another. In the presence of one species, the usual intestinal habitat is not occupied by the other, even though that portion of the spiral valve is available (Simmons and Laurie 1972).

LARVAL BIOLOGY

Ruszkowski (1932), Lynch (1945), and Manter (1951) have described the larvae of three gyrocotylidean species in varying degrees
of detail. The most detailed description is, however, Cole's (1968)
description of the decacanth larva (sometimes termed a "lycophore")
of *G. fimbriata*, and many of her observations are reproduced here by
permission.

Encapsulated embryos of *G. fimbriata* removed from the uterus
require about 19 to 22 days in seawater at 10 to 13C to develop fully.
(*G. parvispinosa* appears to develop a couple of days sooner, but this
requires more critical comparison.) Capsules are apparently sticky
when deposited and soon adhere to the bottom of a glass or plastic
container. Swimming larvae emerge spontaneously over a period of
several days, possibly indicating that not all early uterine embryos
are at identical stages of development. Detailed studies of embryogenesis have not been made, but it is known that the larval hooks
arise in special cells ("oncoblasts"), with the curved blade and then
the shaft being formed (Spencer 1889; Cole 1968), just as in monogeneans and cestodes (Llewellyn 1963; Rybicka 1966).

Newly hatched larvae are approximately 150 μ in length, but are
capable of elongation and contraction within a range of about 100 to
200 μ, and they swim much like *Paramecium*, in a spiral directional
path with the hooks at the trailing (posterior) end. As both Lynch
(1945) and Cole (1968) reported, living larvae are exceedingly difficult to study in detail; fixation in glutaraldehyde or in Bouin's
fluid usually results in about one-third contraction.

Among the more obvious structures, seen even in living larvae,
are two pairs of ducts which open at or very near the anterior end
of the larva (Fig. 1). (There is another, unpaired structure in this
vicinity which may be glandular, but no duct has been seen.) The nar-
rower, more dorsal of these glandular ducts is strongly eosinophilic,
but the underlying, more globose pair are basophilic (Figs. 2-5).
Lynch (1945) states, "They can be traced posteriorly about two-thirds
the length of the larva, but become indistinct and fade from view
just anterior to a mass of vaguely outlined cells located in front of
the larval hooks". This group of about five to eight large cells
(Fig. 1) are presumably the cell bodies of the glands (Cole 1968) and
in living specimens, whole preparations and sections (Fig. 4) appear
to be connected with the ducts which, however, have become quite dif-
fuse in this region.

A cerebral ganglion is located about one-third the body length
from the anterior end (Fig. 1). Roughly rectangular in both transverse and frontal sections (Figs. 3, 4), it has two pairs of major
processes, one anterior, the other posterior. Fibers containing
numerous vesicles of several types are closely packed in the ganglion
and processes, and individual fibers containing small vesicles (20 to
55 mμ diameter) are occasionally seen to pass through muscles into
the epithelium of the larva (Fig. 11), terminating in a manner

similar to the nerve endings observed in the tegument of an adult cestode (Morseth 1966).

Anterior to the cerebral ganglion both pairs of glandular ducts are dorsal in position (Fig. 1, a-a; Fig. 2). In the region of the ganglion (Fig. 1, b-b) they dip sharply ventrad (Fig. 5) in order to

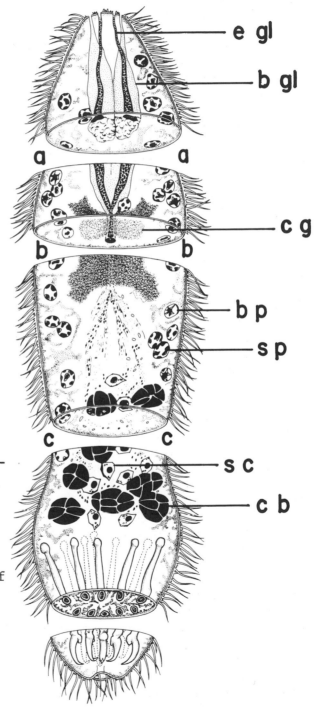

Figure 1

Diagrammatic representation of the larva of *Gyrocotyle fimbriata*. (After Fig. 1; Cole 1968.)

a-a = transverse section anterior to the cerebral ganglion

b-b = transverse section through the cere- bral ganglion

c-c = transverse section through the region of presumptive secretory cells and clustered bodies associated with them

e gl = ducts of eosinophi- lic glands

b gl = ducts of basophilic glands

c g = cerebral ganglion

b p = nucleus of body paren- chymal cell

s p = nucleus of subepi- thelial parenchymal cell

s c = presumptive secre- tory cell

c b = clustered body, one of a group associated with the secretory cells

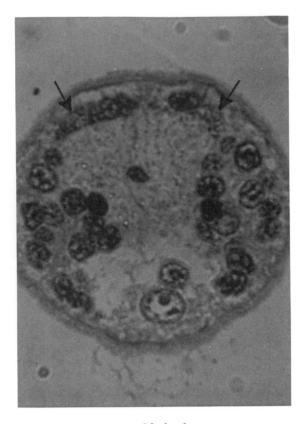

Figure 2. Glandular ducts of *G. fimbriata* larva. Transverse section
anterior to the cerebral ganglion (approximately a-2 of
Fig. 1). The ducts filled with eosinophilic material
(arrows) lie dorsolateral to the ducts containing basophilic secretion. The nucleus of a subepithelial parenchymal
cell is seen midventrally. Bouin's fixative; 1 μ section;
acid fuchsin-aniline blue stain. X3,300. (From Fig. 16,
Cole 1968.)

pass under the ganglion, and in so doing become greatly reduced in
size (Fig. 3). They retain, however, their relative position to one
another (i.e., eosinophilic ducts dorsal).

Seven clustered bodies, each made up of a group of closely packed, darkly stained subunits (Fig. 1, c-c; Fig. 6), are closely associated with the presumed secretory cells, and the three largest, most
medial of these bodies at places appear to be contiguous with the
cells. The function of these clustered bodies cannot confidently be
postulated, and ontogenetic evidence may be required to reveal their
true nature.

Cole (1968) described in detail four types of granules which she
discerned with electron microscopy and referred to them as dark,
striated, gray, and light granules (Fig. 7). The latter two types
could not be associated confidently with specific structures, but
she provided evidence that the dark granules are probably associated
with the basophilic glands and ducts and the striated granules with

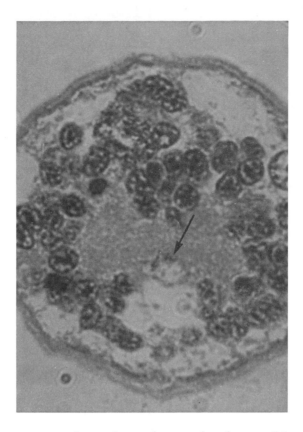

Figure 3. Transverse section through cerebral ganglion of *G. fimbriata* larva (approximately b-b of Fig. 1). The glandular
ducts are greatly narrowed, particularly the dorsal-most
ones (arrow), and lie immediately ventral to the ganglion.
Bouin's fixative; 1 μ section; acid fuchsin-aniline blue
stain. X3,300. (From Fig. 17, Cole 1968.)

the eosinophilic ones. The striated granules were seen in an organized state posterior to the cerebral ganglion. They are highly
organized in the ducts which exit through the epithelium (Fig. 8).
Somewhat posterior to their apertures, the ducts enclosing these
granules are interesting in that no cytoplasmic matrix can be discerned in their inner compartment (Fig. 9). Even more posteriorly,
the striated granules appear to be formed from stacks of membranes
(Fig. 10).
    The epithelium of the larva, which under light microscopy appears
to be anucleate and noncellular, can be seen to be limited externally
by a plasma membrane with the electron microscope (Fig. 11). Typical
cilia project from the surface, and structures between these might be
interpreted as microvilli, although they possibly are cilia cut at
the edge. The ciliary rootlets are about 2 μ in length and lie at a
very acute angle to the surface, an orientation of unknown significance. This condition is also seen in other flatworms, in the nema-
togen, rhombogen, and infusoriform stages of Dicyemida (Bresciani
and Fenchel 1965, 1967; Ridley 1968), in endostyle cilia of tunicates

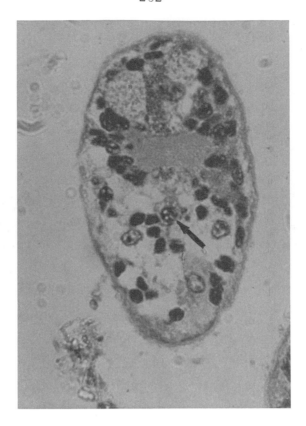

Figure 4. Frontal section of *G. fimbriata* larva. Cerebral ganglion
and glandular ducts are easily discerned. The arrow points
to a pressured secretory cell with which the ducts appear
to connect immediately posterior to the ganglion. Note the
large, numerous, unstained lipid droplets posteriorly.
Bouin's fixative; 1 μ section; acid fuchsin-aniline blue
stain. X2,375. (From Fig. 20, Cole 1968.)

(Olsson 1962), as well as in ciliate protozoans. Internally, the
epithelium is limited by an intercellular space (approximately 50 mμ
thick) which contains a membrane like structure in its center (Fig.
11).

The larva of *G. fimbriata* has two distinctive types of parenchymal cells, each with a distinctive type of nucleus (Figs. 1, 5). The
parenchymal cells, which lie most commonly beneath the superficial
musculature, have the larger (about 3 μ diameter) nuclei which contain scattered, darkly staining chromatin. These cells, which occasionally send processes into the epithelium through the muscle layers, contain large numbers of of granules and extensive areas of
endoplasmic reticulum with numerous associated ribosomes and sparsely
distributed lipid droplets and mitochondria (Fig. 12). The bulk of
the body of the larva is filled with parenchymal cells having smaller
(about 2 μ diameter) nuclei containing a stellate chromatin mass.
Such nuclei are seen to be packed closely, especially in the lateral
areas (Fig. 4). Lipid droplets are particularly concentrated in the

central region of the larvae and are probably inclusions of this second type of cell.

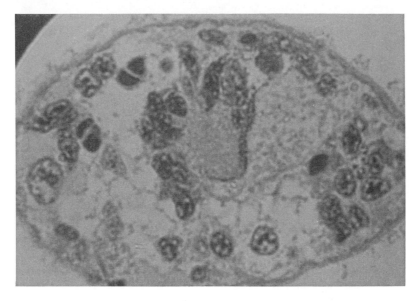

Figure 5.   Larva of *G. fimbriata* showing glandular ducts passing ventrad to the cerebral ganglion. Sagittal section. Bouin's fixative; 1 μ section; acid fuchsin-aniline blue stain. X3,300. (From Fig. 18a, Cole 1968.)

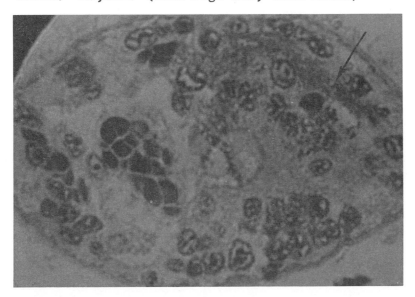

Figure 6. Sagittal section of *G. fimbriata* larva. The most obvious structure is one of the clustered bodies located posteriorly to the cerebral ganglion. The arrow points to an anterodorsal, unpaired, ductless, possibly glandular structure. Bouin's fixative; 1 μ section; acid fuchsin-aniline blue stain. X3,300. (From Fig. 18b, Cole 1968.)

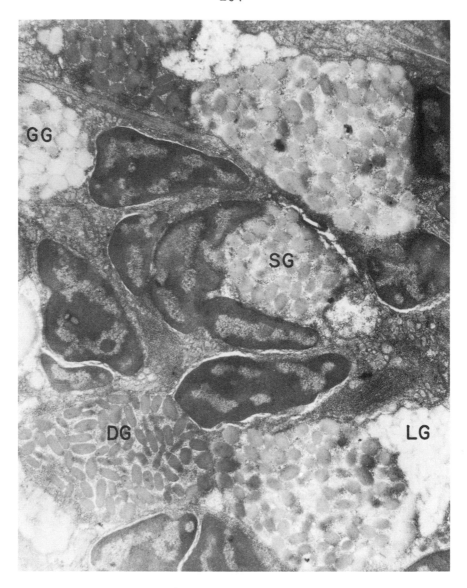

Figure 7. Secretory granules of the glands of the larva of *G. fimbriata*. All four types which have been observed are represented: gray granules (GG); dark granules (DG); light granules (LG); striated granules (SG). X24,000. (From Fig. 21, Cole 1968.)

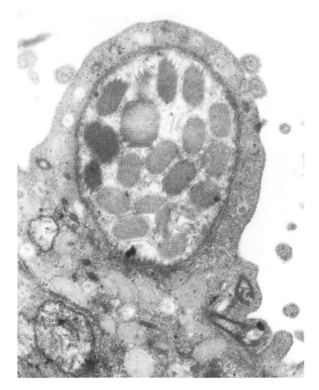

Figure 8.   Duct containing striated granules exiting through the epithelium of *G. fimbriata*.   X82,000.   (From Fig. 24, Cole 1968.)

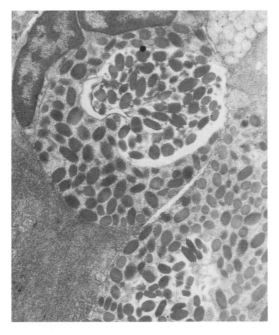

Figure 9. Striated granules in bicompartmented ducts.   X30,100. (From Fig. 25a, Cole 1968.)

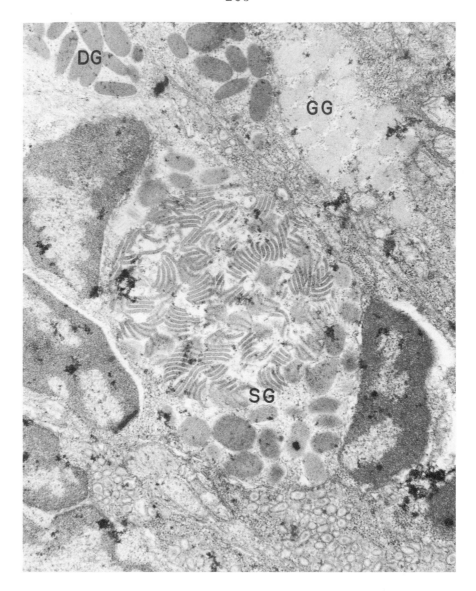

Figure 10. Striated granules apparently being formed in the inner
compartment of a bicompartmented duct. Notation as in
Fig. 7. X34,300. (From Fig. 25b, Cole 1968.)

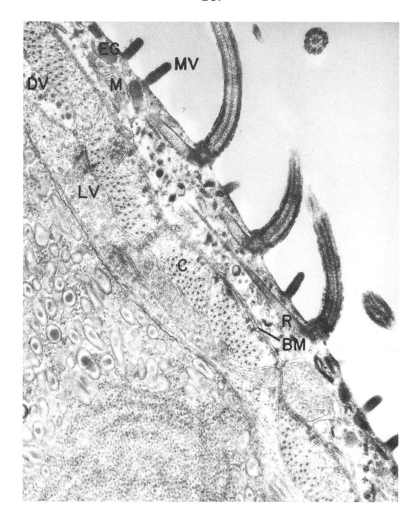

Figure 11. Epithelium of the larva of *G. fimbriata*. EG = granule of
the epithelium; LV = small, light vesicle of nerve fiber;
DV = larger, darker vesicles of nerve fiber; M = mitochondrion; BM = basement membrane; C = circular muscle fibers;
R = ciliary rootlets. Cilia project from the surface,
together with (possibly) microvilli (MV). X53,500.
(From Fig. 6, Cole 1968.) All electron micrographs are of
material fixed with glutaraldehyde-osmium and stained with
uranyl acetate-lead citrate.

208

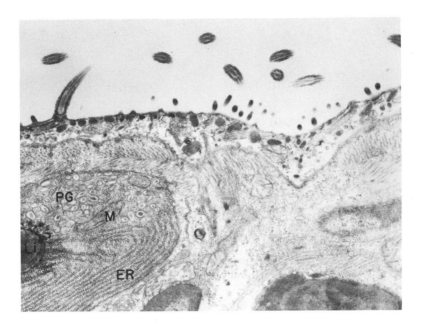

Figure 12. Subepithelial parenchymal cell of the larva of *G. fim-*
briata. PG = granules characteristic of this type of
cell; M = mitochondrion; Li = lipid droplet; ET = endoplasmic reticulum with associated microsomes. X31,700.
(From Fig. 7, Cole 1968).

PARENCHYMAL POSTLARVAE

Perhaps the most singular and curious feature of gyrocotylidean
biology is that postlarval forms in various stages of development
are found embedded in the parenchyma of specimens taken from the val-
vular intestine of host animals. The most commonly occurring embedded forms are hardly larger than the free-swimming larva (Fig. 13,
14). These may occur almost anywhere in the parenchyma, including
that of the holdfast, and there is no evidence of tissue reaction on
the part of the adult nurse animal (Figs. 14, 15). Although usually
only a few postlarvae are to be found in a single host animal, up to
101 of such embedded forms have been reported from *G. fimbriata*
(Lynch 1945). Occasionally, embedded forms in a more advanced state
of development are encountered in the parenchyma (Fig. 15), and in
three instances juveniles have been found in the uterine sac of the
host worm (Fig. 16) (Lynch 1945; Cole 1968). It is to be emphasized
that embedded postlarvae can be found in the tissues of *Gyrocotyle*
that have never reached female maturity and therefore cannot have
produced eggs; indeed, they have been found in quite small (about
2 mm) specimens which are yet juvenile, not having attained male
maturity.

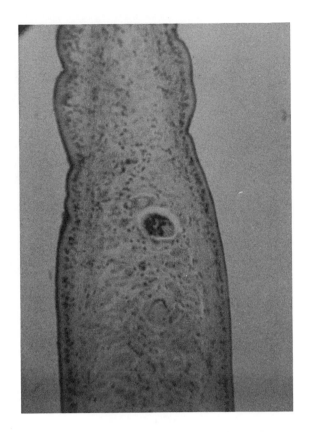

Figure 13
Gyrocotyle postlarva embedded
in parenchyma of *G. fimbriata*.
Harris H & E. X260. (From
Fig. 40a, Cole 1968.)

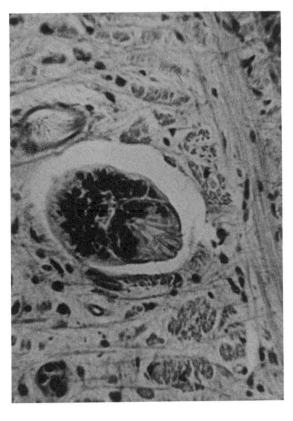

Figure 14
Enlargement of Figure 13
showing definition of hooks
of embedded postlarva. Harris
H & E. X964. (From Fig.
40b, Cole 1968.)

210

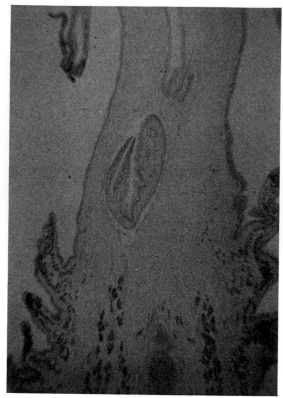

Figure 15
Juvenile gyrocotylidean
embedded in posterior paren-
chyma of *G. fimbriata*.
Harris H & E. X70. (From
Fig. 41, Cole 1968.)

Figure 16
Gyrocotyle parvispinosa with
juvenile in uterine sac.
Nurse worm is male-mature.
Photograph of preparation by
J. E. Lynch, 1944. Galigher's
modification of Harris hema-
toxylin. X12.7. (From
Fig. 42, Cole 1968.)

DISCUSSION

In the brief outline above, attention has been directed to some of the interesting and ill understood attributes of *Gyrocotyle* and gyrocotylidean infections: (1) speciation into two major complexes, each with a representative in a given host species; (2) infection intensity, which is apparently regulated toward a pair of individuals; (3) larval biology, which is almost totally unexamined; and (4) the presence of postlarval *Gyrocotyle* in the tissues of adult worms removed from the intestine.

With regard to the latter phenomenon, Fuhrmann (1930) considered the presence of postlarvae in the parenchyma to be accidental. He suggested that they enter *Gyrocotyle* incidental to penetration of the intestinal mucosa where, he thought, they may undergo early post-larval development. There is no positive evidence to support this, however, and Lynch (1945) has suggested that the considerable frequency with which such developing postlarvae are encountered in the tissues of *Gyrocotyle* may indicate that this is not an accidental occurrence but a normal phase in the life history of the group. We (Simmons and Laurie 1972) have noted several means by which *Gyrocotyle* adults may be liberated from the intestines of host fish and, as the worms are long-lived in seawater, we have suggested that the rare "massive" infections may result from the ingestion of such liberated worms containing large numbers of parenchymal postlarval forms. Whether or not this frank speculation proves true, clearly we cannot fully understand the biology of *Gyrocotyle* until the meaning of these postlarvae is determined with certainty. Under favorable conditions it is possible to procure and husband chimaeroid fish, at least *H. colliei*, and thus transmission experiments are not out of the question.

In addition, a great deal needs to be learned about the biology of the free-swimming larvae. As was pointed out above, there is still a measure of uncertainty regarding the manner in which infections become established initially in the small fish, and whether or not they begin as infections with numerous individuals. If this were so, it would suggest utilization of some intermediate host, for this would be perhaps the most likely means of concentration of these dispersal forms (but not the only means, of course; the eggs, which are sticky, might become affixed to a substratum en masse and the larvae, upon hatching, remain in close proximity for a considerable period). However, several authors (Manter 1951; Llewellyn 1970) believe that the life history of gyrocotylideans may prove to be a direct one. The larva is almost certainly one which is capable of tissue penetration. The mere presence of postlarvae in the tissues of immature worms is sufficient to justify such a conclusion. Moreover, Manter (1951) found that larvae of *G. rugosa* would readily penetrate the cut edge (but not the intact surface) of a piece of host intestinal mucosa or a mass of mucus from the intestine. Following penetration, the staining properties of one pair of the larval glands, perhaps the dorsal pair, became altered.

Certainly, an analysis of factors involved in the attraction of larvae, in the stimulation of penetration, and in the presumed further development in tissue would be a valid line of investigation. Until evidence to the contrary is available, it might be assumed that some degree of development in a tissue site precedes entry into the gut of the fish host. It is appropriate here to mention some unpublished observations of the author. The small serpulid annelid, *Mercierella enigmatica* Fauvel, 1923, readily filters *G. fimbriata* lycophores from the surrounding water; one day later, however, the larvae have been digested, and their hooks are found scattered in food boli in the annelid's intestine. Similarly, the splash pool harpacticoid copepod, *Tigriopus californicus* (Baker, 1912), avidly ingests *G. fimbriata* lycophores; later, hook coronae are obvious in the fecal pellets of the copepods.

In conclusion, we should consider the interest directed toward gyrocotylideans in terms of their systemic position and phylogenetic relationships. Bychowsky (1937) and Lynch (1945) compared gyrocotylideans with monogeneans and with cestodes and independently pointed out the similarities between *Gyrocotyle* and the monogeneans. The former author, guided by an earlier suggestion of Spengel (1905) that there was a direct relationship between monogeneans and cestodes, discussed the issue at length and placed particular emphasis upon gyrocotylideans, which, he concluded, should be accorded an intermediary position between Monogenoidea and Cestoidea. Bychowsky reiterated this view in 1957 and cited a number of Soviet parasitologists who had accepted it. Elsewhere, however, Bychowsky's views have not been widely accepted; one of the principal proponents of the theory of interrelationships of monogeneans and cestodes is Llewellyn (1963) who lucidly detailed his reasons. Subsequently, Llewellyn (1965) used this theory as a point of departure for a most stimulating discussion of the possible routes of evolution of parasitic platyhelminths.

Certainly, *Gyrocotyle* resembles cestodes in certain ways, and it is principally because of the lack of a digestive tract that it has been classified as a cestoid. Lyons (1969) provided an excellent description of the fine structure of the body wall of *G. urna* and demonstrated that the covering layer resembled that of cestodes more than it did that of monogeneans. Cole (1968) examined the fine structure of *G. fimbriata*, and her description differs little, if any, from that of Lyons (Figs. 17-20). Both of these workers point out that the kind of epidermis found in *Gyrocotyle* and the tapeworms is a result of endoparasitism involving loss of the gut, with the consequence that the epidermis has become specialized. Thus, the similarity is most likely of functional, and not phylogenetic, significance.

It is interesting to compare the surface elaboration of *Amphilina* (Fig. 21) with that of *Gyrocotyle* (Fig. 17). That of the latter is developed into numerous microvilli, as in eucestodes, but differs in lacking electron-dense tips (Lyons 1969; Cole 1968). The general

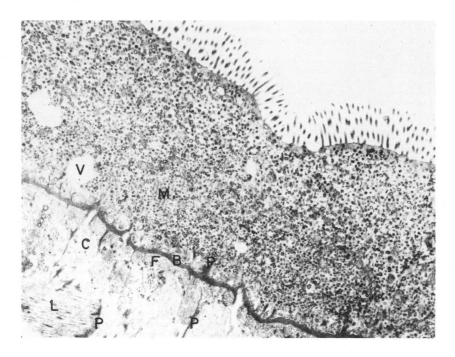

Figure 17. Tegument of *G. fimbriata* adult. M = mitochondrion; V = large vesicle; B = basement layer; P = processes of basement layer; F = fibrous layer; C = circular muscle layer; L = longitudinal muscle layer. X24,600. (From Fig. 30, Cole 1968.)

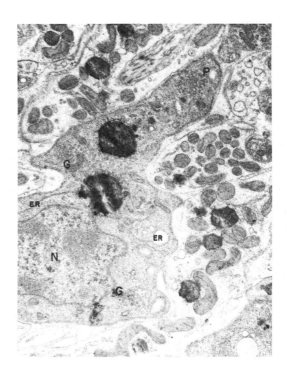

Figure 18
Subtegumental cells of *G. fimbriata* adult. Note light cells containing numerous mitochondria, and the darkly stained cell with a basal nucleus (N) and a process (P) extending toward the tegument. G = Golgi bodies; ER = elongate and circular endoplasmic reticulum. X35,500. (From Fig. 34, Cole 1968.)

Figure 19
Subtegumental cells near the
tegument of *G. fimbriata* adult.
Several vesicles (V) forming
at the Golgi bodies (G) resemble certain granules in the
tegument.   B = basement layer;
F = fibrous layer; C = circular muscle layer; L = longitudinal muscle layer.   X33,400.
(From Fig. 35, Cole 1968.)

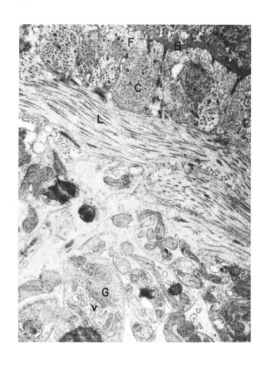

Figure 20
Parenchymal cells of *G.
fimbriate* adult.   Note the
intercellular collagen like
fibers (F).   X42,700. (From
Fig. 33, Cole 1968.)

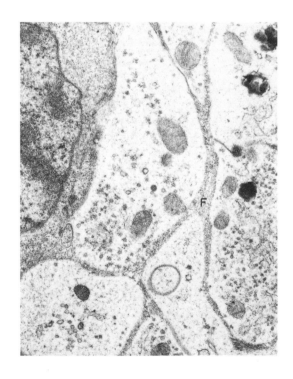

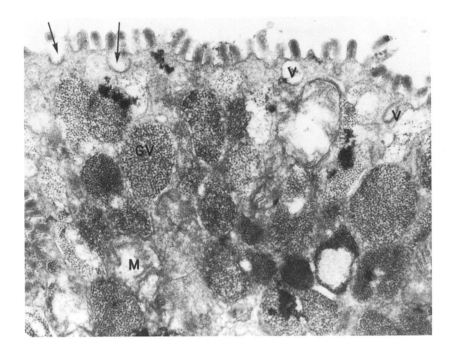

Figure 21. Tegument of *Amphilina bipunctata*. Note the clear vesicles
(V) just beneath the outer surface, and the depressions
(arrows) formed at the surface. M = mitochondrion;
GV = granular vesicles which occur in the tegument.
X72,900. (From Fig. 38, Cole 1968.)

tegumental structures of *Amphilina* conform to the general pattern
exhibited by parasitic flatworms except that no basement membrane
has been observed (Cole 1968). However, there are more sparse and
greatly reduced microvilli, large granules (apparently all of a single type), and abundant vesicular associations with the surface, in
contrast to the condition in *Gyrocotyle*. This species, *Amphilina
bipunctata*, occurs in the body cavity of white sturgeons (*Scipenser
transmontanus*) and green sturgeons (*A. medirostris*). Pinter (1905)
suggested that *Amphilina* represented a neotenic eucestode larva.
Janicki concurred, and his life history studies (Janicki 1928a, b)
indicated the pseudophyllodoid nature of the worm. Bychowsky (1937)
argued for a separation of the amphilinans and gyrocotylideans, and
more recently others (Dubinina 1960; Llewellyn 1965; and Lyons 1966)
provided several lines of evidence which would be compatible with a
dissociation of these groups of so-called Cestodaria.

Progress toward a more complete understanding of *Gyrocotyle* probably will not proceed rapidly, in part because the host animals are
easily collected and studied at relatively few locations and also
they are not the easiest animals with which to work. However, the
author has considerable confidence that although "nature relinquishes
her secrets with the greatest reluctance," in time sufficient sub-
stantive information will be brought to bear on the many unsolved
aspects of gyrocotylidean biology.

ACKNOWLEDGMENTS

For several summers the staff of the University of Washington
Friday Harbor Laboratories provided unmatched cooperation and support
for studies with gyrocotylideans. Our gratitude is likewise unmatch-
ed. We thank Mr. Arthur Oakley and Mr. Jim Galbreath of the Oregon
Fish Commission Columbia River Investigations for making possible
collection of sturgeon for N. C. Cole's study. Miss Cole, now Mrs.
N. C. Melcerek, has provided a very fine description of *Gyrocotyle*
larvae, as will be apparent to the reader. Mrs. Kirstin Clark
Nichols helped the author with photographic printing. Mrs. Marcia
Kier made the preliminary drawing of the larva (Fig. 1), and Mrs.
Celeste Green very kindly made the finished illustration.

LITERATURE CITED

Bresciani, J., and Fenchel, T. 1965. Studies on dicyemid Mesozoa.
    I. The fine structure of the adult (the nematogen and rhobogen
    stage). Videnskabelige medd. Dansk naturhist. Foren., Copenha-
    gen 128:85-92.

_____, and _____. 1967. Studies on dicyemid Mesozoa. II. The fine
    structure of the infusoriform larva. Ophelia 4:1-18.
Bychowsky, B. E. 1937. [Ontogenesis and phylogenetic interrelation-
    ships of parasitic flatworms] (in Russian, German summary).
    Izvest. Akad. Nauk. S.S.S.R., Ser. Biol. 4:1353-83.
_____. 1957. [Monogenetic trematodes: their systematics and phy-
    logeny] Akad. Nauk S.S.S.R., Moscow. (A.I.B.S. translation,
    1961, W. J. Hargis, Jr., ed.)
Cole, N. C. 1968. Morphological studies on Cestodaria. M. A.
    thesis, University of California.
Dienske, H. 1968. A survey of the metazoan parasites of the rabbit-
    fish, *Chimaera monstrosa* L. (Holocephali). Netherl. J. of Sea
    Research 4:32-58.
Dubinina, M. N. 1960. [The morphology of Amphilinidae (Cestodaria)
    in relation to their position in the system of flatworms]
    Doklady Akad. Nauk S.S.S.R. 135:501-504. (A.I.B.S. translation,
    pp. 943-45.)
Fuhrmann, O. 1930. Erste Unterklasse der Cestoidea: Cestodaria
    Monticelli. In Handbuch der Zoologie, ed. W. Kükenthal and T.
    Krumbach, Vol. 2, Pt. 2, pp. 144-80. Berlin: Walter de Gruyter.
Halvorsen, O., and Williams, H. H. 1968. Studies on the helminth
    fauna of Norway. IX. *Gyrocotyle* (Platyhelminthes) in *Chimaera
    monstrosa* from Oslo Fjord, with emphasis on its mode of attach-
    ment and a regulation in the degree of infection. Nytt Mag.
    Zool. 15:130-42.
Janicki, C. 1928a. Über die Lebensgeschichte von *Amphilina foliacea*,
    dem Parasiten des Wolga-Sterlets, nach Beobachtungen und Experi-
    menten. Naturwissenschaften 16:820-21.

_____. 1928b. Die Legensgeschichte von *Amphilina foliacea* G. Wagen., Parasiten des Wolga-Sterlets, nach Beobachtungen und Experimenten. Arb. Biol. Wolga-Station, Saratow 10:101-34.

Land, J. van der, and Dienske, H. 1968. Two new species of *Gyrocotyle* (Monogenea) from chimaerids (Holocephali). Zool. Mededel., Leiden 43:97-105.

_____, and Templeman, W. 1968. Two new species of *Gyrocotyle* (Monogenea) from *Hydrolagus affinis* (Brito Capello) (Holocephali). J. Fish. Res. Bd. Canada 25:2365-85.

Laurie, J. S. 1971. Carbohydrate absorption by *Gyrocotyle fimbriata* and *Gyrocotyle parvispinosa* (Platyhelminthes). Exp. Parasitol. 29:375-85.

Llewellyn, J. 1963. Larvae and larval development of monogeneans. Advances in Parasitology 1:287-326.

_____. 1965. The evolution of parasitic platyhelminths. In Evolution of parasites, ed. A. E. R. Taylor, Third Symposium of the British Society for Parasitology, pp. 47-78. Oxford:Blackwell.

_____. 1970. Monogenea. J. Parasitol. 56 (Section II, Pt. 3): 493-504.

Lynch, J. E. 1945. Redescription of the species of *Gyrocotyle* from the ratfish, *Hydrolagus colliei* (Lay and Bennet), with notes on the morphology and taxonomy of the genus. J. Parasitol. 31: 418-46.

Lyons, K. M. 1966. The chemical nature and evolutionary significance of monogenean attachment sclerites. Parasitology 56: 63-100.

_____. 1969. The fine structure of the body wall of *Gyrocotyle urna*. Zeit. Parasitenk. 33:95-109.

Manter, H. W. 1951. Studies on *Gyrocotyle rugosa* Diesing, 1850, a cestodarian parasite of the elephant fish, *Callorhynchus milii*. Zool. Publ. Victoria Univ. Coll., Wellington 17:1-11.

Morseth, D. J. 1966. The fine structure of the tegument of adult *Echinococcus granulosus, Taenia hydatigena,* and *Taenia pisiformis*. J. Parasitol. 52:1074-85.

Olsson, R. 1962. The relationship between ciliary rootlets and other cell structures. J. Cell Biol. 15:596-99.

Pinter, T. 1905. Über *Amphilina*. Verhandl. Ges. deut. Naturforscher u. Ärzte 77:196-98.

Ridley, R. K. 1968. Electron microscopic studies on dicyemid Mesozoa. I. Vermiform stages. J. Parasitol. 54:975-98.

Ruszkowski, J. S. 1932. Etudes sur le cycle evolutif et sur la structure des Cestodes de mer. II. Sur les larves de *Gyrocotyle urna* (Gr. et Wagen.). Bull. intern. acad. polon. sci., Classe sci. math. nat., Ser. B, 11:629-41.

Rybicka, K. 1966. Embryogenesis in cestodes. Advances in Para.- sitology 4:107-86.

Simmons, J. E., and Laurie, J. S. 1972. A study of *Gyrocotyle* in the San Juan Archipelago, Puget Sound, U.S.A., with observations on the host, *Hydrolagus colliei* (Lay and Bennett). Intern. J. Parasitol. 2:59-77.

_____, Buteau, G. H., Jr., Macinnis, A. J., and Kilejian, A. 1972. Characterization and hybridization of DNAs of gyrocotylidean parasites of chimaeroid fishes. Intern. J. Parasitol. 2:273-78.

Spencer, W. B. 1889. The anatomy of *Amphiptyches urna* (Grube and Wagener). Trans. Roy. Soc. Victoria 1:138-51.

Spengel, J. W. 1905. Die Monozootie der Cestoden. Z. wiss. Zool. 82:252-87.

Ward, H. B. 1912. Some points on the general anatomy of *Gyrocotyle*. Zool. Jahrb., Suppl. 15 Festschr. 60 Geburtstag, J. W. Spengel, pp. 717-38.

Wardle, R. A. 1932. The Cestoda of Canadian fishes. I. The Pacific coast region. Contr. of the Canad. Biol. Fish., 7:223-43.

_____, and McLeod, J. A. 1952. Biology of tapeworms. Minneapolis: Univ. of Minnesota Press.

Watson, E. E. 1911. The genus *Gyrocotyle*, and its significance for problems of cestode structure and phylogeny. Univ. California Publ. Zool., Berkeley 6:353-468.

Community structure among the animals inhabiting the coral Pocillopora damicornis at Heron Island, Australia

Wendell K. Patton
Ohio Wesleyan University
Delaware, Ohio

The branching coral, *Pocillopora damicornis* (L), is one of the best known of all reef coral species, ranging from the Indian Ocean across the Pacific to the Gulf of Panama and occurring commonly in relatively shallow waters. Colonies of this coral are abundant on the reef flat at Heron Island, in the Capricorn group off the coast of Queensland, Australia, where they harbor a variety of macroscopic, freely mobile animal species. This paper deals with this community of organisms and with some of the adaptations and possible interrelationships shown by the most characteristic species. It is based on work done at Heron Island at various times during the periods 1956–59 and 1970–71.

METHODS

Collections were made primarily from the Heron Island reef with the rest from the adjacent Wistari Reef. Most collecting was done on the reef flat during the daytime low tide, when the corals were covered by only small amounts of water. A wooden frame containing brass screen of ten meshes to the inch was placed beside the colony to be collected. The coral then was quickly pried loose into the screen with a small wrecking bar and the screen lifted from the water and placed on a small wooden table. The dead coral was broken off and discarded, the living portions broken apart, and the animals inhabiting them collected. This method caught the small shrimp and other species that leave the colony during the collection process. Specimens smaller than the smallest egg-bearing female of a species were considered to be juveniles. The sex of shrimp specimens was determined by looking for the appendix masculina on the second pleopod.

During the periods I was at Heron Island, it was unfortunately not possible to keep corals or their inhabitants alive in the laboratory for more than brief periods of time. Some observations of the behavior of living animals were made, however, and a few specimens survived the journey back to Brisbane and so permitted longer observation. Feeding habits were investigated by removing the stomach from a preserved specimen and placing it in 95% alcohol where adhering tissues were removed. In some cases the contents were stained, while in others the stomach was opened directly in Diaphane mounting medium (Will Scientific, Rochester, New York), covered, and dried before examination. The species diversity of each collection was computed from Brillouin's formula,

$$H = \frac{1}{N} \log \frac{N!}{N_1! N_2! \dots N_s!}$$

which Pielou (1966) has shown to be the appropriate information-theoretical formula for use on collections in which each individual has been identified and counted. The evenness with which the individuals were distributed among the species present is given by the formula

$$H_{max} = \frac{1}{N} \log \frac{N!}{\left\{\left[\frac{N}{s}\right]!\right\}^{s-r} \left\{\left(\left[\frac{N}{s}\right]+1\right)!\right\}^{r}}$$

where [N/s] is the integer part of N/s and r = N - s[N/s] (Pielou 1966).

THE CORAL HOST

The planula larvae of *P. damicornis* settle on submerged hard substrates on the reef flat and the resulting colonies grow outward in a spherical or conical form. The coral polyps are small and are often expanded in the daytime. Most colonies have a light fawn coloration, although some have a definite pinkish cast. As a colony gets larger, the tissue covering the innermost branches generally dies and the exposed skeleton is overgrown by algae, sponges, and other encrusting organisms which may harbor an extensive fauna quite different from that of the living coral.

THE CORAL ASSOCIATES

Animals were collected from a large number of *Pocillopora* colonies. Data for thirty-eight representative collections are given in
Table 1. Figure 1 gives an idea of the spatial relationship of the
coral to its inhabitants, while some of the more common associates
are shown in Figure 2.

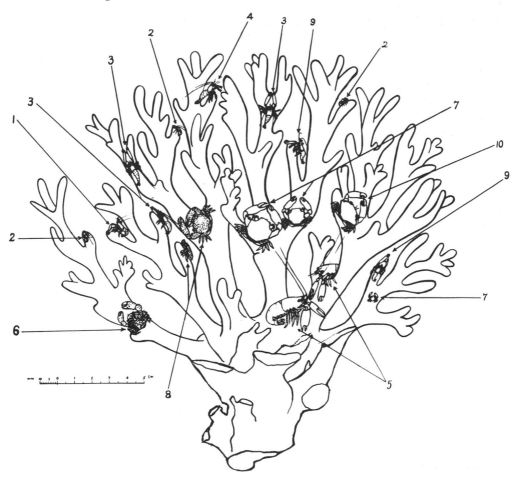

Figure 1. Diagrammatic cross section of a *Pocillopora damicornis*
colony showing some of the typical associates
1. *Periclimenes amymone*
2. *P. madreporae*
3. *Harpiliopsis beaupressi*
4. *Thor amboinensis*
5. *Alpheus lottini*
6. *Cymo andreossyi*
7. *Trapezia cymodoce*
8. *T. ferruginea* form *areolata*
9. *Periclimenes spiniferus*
10. *Chlorodiella nigra*

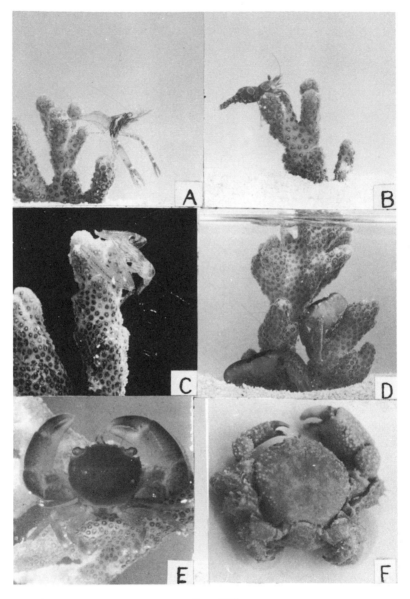

Figure 2. Representative *Pocillopora* associates

    A. *Periclimenes spiniferus*
    B. *Thor amboinensis*
    C. *Harpiliopsis beaupressi*
    D. *Alpheus lottini*
    E. *Trapezia cymodoce*
    F. *Cymo andreossyi*

TABLE 1
Fauna of *Pocillopora damicornis**

COLONY SIZE	Small – diameter 8-15 cm											
COLLECTION NUMBER	309	333	341	342	355	356	Av.	304	306	307	311	314
Agile shrimp												
Periclimenes amymone	1	2										
Periclimenes madreporae								2M,2F			1M,1F	3
Periclimenes spiniferus		1M	1F									1
Thor amboinensis								3F,1J				5
Sedentary shrimp												
Periclimenes sp.										1F	1M,1F	1M,1F
Harpiliopsis beaupressi	1M,1F	1M,1F		1F				1M,1F	1M,2F			
Alpheus ventrosus	1M,1F	1M,1F	1M,1F	1M,1F	1M,1F	1M,1F		1M				
Crabs												
Chlorodiella nigra	1J	1M			1J			1M	12J	1J		1J
Phymodius ungulatus										1J	1J	
Cymo andreossyi	1M,1F							1M	12J		1M,1F	1M
Domecia hispida									1M,1F			
Trapezia cymodoce	3J		1F	1J				1M,**1F**,2J	3J	1M,1F	2M,2F	1M,1F
Trapezia ferruginea	1M,1F	1M,1F	1M,1F	1M,1F	1M,1F					1M,1F		
Trapezia ferruginea form *areolata*						1M,1F, 1J		1F	1J		1M,1F, 2J	
Trapezia ferruginea form *guttata*												
Fish												
Paragobiodon echinocephalus				1				1				2
No. individuals per colony	13	10	6	7	5	5	7.7	19	33	7	15	16
No. species per colony	7	6	4	5	3	2	4.5	9	6	5	6	8
Diversity (H)	.59	.54	.38	.44	.30	.20	.41	.66	.54	.44	.56	.62
Evenness (J)	.96	1.00	1.00	1.00	1.00	1.00	.99	.91	.80	1.00	.95	.90

*Colonies arranged according to size. Where symbionts sexed M=male, F=female, J=juvenile; 300 series=parts of Heron Island Reef, December 1957; 500 series=Heron Island Reef, August 1958; W series=Wistari Reef, December 1957.

TABLE 1 (Cont.)

COLONY SIZE	Medium – diameter 15–25 cm										
COLLECTION NUMBER	315	332	334	338	345	349	354	360	512	W8	W9
Agile shrimp											
Periclimenes amymone		4M,4F					2M,1F	11			
Periclimenes madreporae	3	1M,1F			7		2			7	11
Periclimenes spiniferus	1M,1F	1M,2F	2M,3F	1F	1M,1F		3M,4F	1M,1F,1J	2M,1F		1F
Thor amboinensis	2M,1F	3	2M,2F	1M,1F	1M,1F						1
Sedentary shrimp											
Periclimenes sp.											
Harpiliopsis beaupressi		1F	1F	1M	1M,2F	1M	2M,2F		3M,4F	1M,2F	2M,2F
Alpheus ventrosus	1M,1F	1M,1F		1F	2	1M	1M,1F	1M,1F	1M	1M,1F	1M,1F
Crabs											
Chlorodiella nigra	1M,1F	6	5	2M,1F,1J	1J	1M,1F	2M,3F		1M	1M,1F	1M,2F
Phymodius ungulatus		1M								1M	
Cymo andreossyi		1M	1M	1	1M	1M,1F	1M	1J	1M,1J	1M	2M
Domecia hispida											1M
Trapezia cymodoce	1M,1F		1M,1F	1M,1F	2F	2M,1F,3J	1M,1F,2J	1M,1F,1J	2M,2F,7J	1F,1J	2M,2F,4J
Trapezia ferruginea				1M	1J	1F					
Trapezia ferruginea form *areolata*		1M,1F		1M,1F,1J	1M,1F			1F		1M,2	1M,2F
Trapezia ferruginea form *guttata*											
Fish											
Paragobiodon echinocephalus											2
No. individuals per colony	14	29	18	16	23	13	28	21	25	21	38
No. species per colony	6	10	6	9	10	6	8	6	6	7	11
Diversity (H)	.58	.73	.55	.66	.71	.49	.70	.49	.51	.58	.75
Evenness (J)	1.00	.89	.89	.94	.91	.84	.94	.77	.79	.88	.86

TABLE 1 (Cont.)

COLONY SIZE	Medium – diameter (cont.)			Large – diameter 25-35 cm					
COLLECTION NUMBER	W16	W25	Av.	317	329	335	353	357	361
Agile shrimp									
Periclimenes amymone							1M		
Periclimenes madreporae	6	2		2	2		17	31	12
Periclimenes spiniferus				8	8	1F	2M,2F	2M,3F	6
Thor amboinensis				6				1M	
Sedentary shrimp									
Periclimenes sp.									
Harpiliopsis breauipressi	2M,1F	1		10	2M,4F	2M,2F	3F	1M	1
Alpheus ventrosus	1M,1F	1M,1F		1M,1F	2	1M	1M,1F	1M,1F	
Crabs									
Chlorodiella nigra	4M,1F,1J	1J		2M,4F	2M,2F,3J	1F	5	3M,9F,1J	3F,1J
Phymodius ungulatus	1M	1F						1M,1F	2M
Cymo andreossyi							1F	1M,1F	
Domecia hispida									
Trapezia cymodoce	1M,2F,2J	1J		1M,1F	1M,1F	1M,2F,2J	1M,1F,3J	2M,3F,3J	2M,2F
Trapezia ferruginea	1M,2F	1M,1F		1M,1F 1M,1J					
Trapezia ferruginea form *areolata*	1M							1F	
Trapezia ferruginea form *guttata*									
Fish									
Paragobiodon echinocephalus	2			1				2	1
	28	11	21	39	27	11	38	68	30
	8	8	7.5	9	6	5	8	11	7
	.70	.61	.60	.72	.59	.44	.62	.66	.59
	.94	1.00	.90	.89	.90	.86	.80	.71	.83

TABLE 1 (Cont.)

COLONY SIZE	Large – diameter (cont.)					Very large – diameter over 35 cm					No. of colonies inhabited	No. of individuals
COLLECTION NUMBER	513	520	W2	W15	Av.	313	316	328	331	Av.		
Agile shrimp												
Periclimenes amymone	11	1F				14					6	26
Periclimenes madreporae	14	1F	4	13		21		20	4		24	181
Periclimenes spiniferus							1M,1F	5M,5F	2M,4F		23	115
Thor amboinensis	2			1M,1F				5			13	40
Sedentary shrimp												
Periclimenes sp.							1M,1F				1	2
Harpiliopsis breaupressi	12	4M,2F	1M,3F	6		19		19M,11F,5J	6M,4F		31	156
Alpheus ventrosus	1M	1M,1F	1M,1F	1M,1F		1M,1F		1M,1F	1M,1F		33	61
Crabs												
Chlorodiella nigra	1F, 2J		1	1M,1F		47	1M,5F	4F	3M,8F		33	167
Phymodius ungulatus	2M										8	11
Cymo andreossyi		2J		1M		1M,2F	1M,2F	2M	2F		23	48
Domecia hispida											2	3
Trapezia cymodoce	1M, 2F, 3J		1M,1F,1J	4J		2M,4F,3J	1M,2F	1M,2F	3M,2F		33	125
Trapezia ferruginea						1M					6	8
Trapezia ferruginea form *areolata*		1M,1F	1M,1F	1M,1F		1M,1F		1M,1F	1J		23	46
Trapezia ferruginea form *guttata*						1M					2	2
Fish												
Paragobiodon echinocephalus				3			1				10	16
	51	14	16	35	33	119	17	83	41	65		
	8	6	6	9	7.5	10	6	9	8	8.3		
	.68	.51	.57	.69	.61	.69	.56	.66	.69	.65		
	.85	.87	.96	.85	.85	.74	.92	.75	.89	.83		

In the majority of cases, I believe that the data in Table 1 represent the total populations of mobile, macroscopic animals inhabiting the various corals at the time of collection. However, an occasional specimen may have been lost and a few of the fish and crabs found lying in the screen may have been discarded as having come from dead coral when in fact they had been inhabiting the living parts of the colony. No fish other than *Paragobiodon echinocephalus* were found on the living parts of the colonies listed, although colonies collected at other times contained juveniles of various species. Microscopic inhabitants of the living coral, such as copepods, or immobile ones such as the date clam, *Lithophaga*, or the gall crab, *Hapalocarcinus*, were not included in this study.

It is apparent from Table 1 that no one species dominates the *Pocillopora* fauna and that on most colonies the modest number of individuals present is distributed fairly evenly among several species. These animals can be divided into four groups: agile shrimp, sedentary shrimp, crabs, and fish. The most important members of each group are listed in Table 2 and will be discussed individually below.

A study of stomach contents (Table 2) indicates a diet of zooxanthellae and coral tissue for the fish *Paragobiodon echinocephalus* but shows a surprising uniformity among the twelve most common crustacean species. Except for one instance of what appeared to be sponge tissue in the stomach of *Chlorodiella nigra*, there is no indication that the crustaceans utilize living organisms as food. There are no remains of phyto- or zooplankton as would have been expected if they were plankton feeders or if they removed material collected by the coral polyps. There are no pieces of animal tissue to indicate a predatory diet and no sign of zooxanthellae or nematocysts to indicate a diet of coral tissue or its products. Instead, the stomachs contain sediment particles and soft amorphous material. The latter often fills the stomach and so it is probably ingested rather than secreted by the crustacean itself. The soft material contains fewer bacteria than are usually found in organic aggregates and thus may consist largely of coral mucus. Some of this material shows metachromasia with toluidine blue as do the mucus cells of coral (Goreau 1956) while other seemingly identical material does not.

Many of the reef-dwelling decapod Crustacea are most active at night and it thus seems possible that the *Pocillopora* crustaceans may leave the coral at night to feed and merely use the coral head as a daytime hiding place. To check this possibility, nine medium colonies were collected during three nighttime low tides. Three of the agile shrimp to be discussed below were rare in comparison with their daytime abundance. All of the remaining species in Table 2 occurred in numbers comparable to those on the medium colonies in Table 1 and presumably do not systematically forsake the coral at night.

TABLE 2
Size and stomach contents of the symbionts collected from at least 6 coral colonies

Species	Size in mm of a typical* adult specimen	No. of stomachs examined	No. of stomachs		Items found in stomachs listed in order of abundance**
			empty	almost empty	
Agile shrimp *Periclimenes amymone*	17	4	-	-	soft material(4); mineral particles(4); a few sponge spicules (4); foraminifera skeletons (2) and empty fungal filaments (1)
Periclimenes madreporae	11	6	2	2	soft material (4); mineral particles (4); sponge spicules (2)
Periclimenes spiniferus	17	5	-	-	large amounts of mineral particles (5) and soft material containing mineral particles (5); small amounts of foraminifera skeletons and fragments (4); sponge spicules (5) and algal filaments (2)
Thor amboinensis	11	5	2	-	soft material (3) with abundant mineral particles (3) and fragments of sponge spicules (1)

TABLE 2 (Cont.)

Species	Size in mm of a typical adult specimen [*]	No. of stomachs examined	No. of stomachs		Items found in stomachs listed in order of abundance [**]
			empty	almost empty	
Sedentary shrimp					
Harpiliopsis beaupressi	14	7	2	2	soft material (5); mineral particles (5); fragments of sponge spicules (2)
Alpheus ventrosus	28	11	–	1	soft material (10); broken and entire sponge spicules (8); mineral particles (11); foraminifera skeletons (3); pieces of crustacean exoskeleton (2); a few algal filaments (2)
Crabs					
Chlorodiella nigra	14	11	1	1	mineral particles (10); soft material (9); entire and fragmented foraminifera tests and sponge spicules (4); algal filaments (2); sponge tissue(1)
Phymodius ungulatus	18	4	–	–	soft material(4);mineral particles (4); foraminifera fragments (1)
Cymo andreossyi	12	6	–	–	large numbers of mineral fragments (6); soft material with inclusions of small particles (6); sponge spicules (2)

TABLE 2 (Cont.)

Species	Size in mm of a typical adult specimen*	No. of stomachs examined	No. of stomachs empty	No. of stomachs almost empty	Items found in stomachs listed in order of abundance**
Trapezia cymodoce	18	10	-	1	soft material (10) with abundant small inclusions of sponge spicule fragments (10); mineral particles (10); foraminifera fragments (3); abundant bacteria (2)
Trapezia ferruginea	13	5	-	2	soft material (5) containing some spicule fragments and mineral particles (5); abundant bacteria (2)
Trapezia ferruginea form *areolata*	11	7	-	2	soft material (7) containing variable amounts of spicules (6); mineral fragments (6); plant debris (1); algal tissue (1) and skeleton of host coral (1)
Fish *Paragobiodon echinocephalus*	15	7	2	1	zooxanthellae (5); soft material (5); coral tissue (2)

*Length for shrimp and fish; carapace width for crabs.
** () = No. of stomachs in which each item was found.

Agile Shrimp

These animals rest lightly on the coral branches rather than holding tightly to them and leave the coral readily when it is disturbed. All have the sharply pointed dactyls on the walking legs which are characteristic of free-living shrimp. *Periclimenes amymone*, *P. madreporae*, and *Thor amboinensis* were represented by a total of only five individuals from nine medium-sized colonies collected at night. They thus may tend to leave the coral at night. The evidence for this is hardly clear-cut, however, since *P. amymone* has a very patchy distribution, while *P. madreporae* and *T. amboinensis* are the two smallest species and individuals could have been lost or overlooked due to the difficulties of nighttime collection.

The three species of *Periclimenes* belong to the subfamily Pontoniinae of the Family Palaemonidae, a group having many species living in association with other organisms. The genus *Periclimenes* however, contains both free-living and associated species. Bruce (1971) calls attention to the fact that the Pontoniinae associated with corals frequently have well-developed rows of curved setae on the ventral margin of the chela of the small first pair of leggs (first pereiopods). He suggests that these setae are associated either with feeding or with cleaning the shrimps from coral mucus. I believe that these setae are an adaptation for collecting the sediment and coral mucus on which the animals feed. In *Anchistus custos*, a shrimp that scrapes up the food-containing mucus produced by its bivalve host, the setae on the ventral borders of the chelae of the first legs are the only important food-collecting structures (Johnson and Liang 1966). The second pair of legs is also chelate and is invariably larger than the first. In the majority of the symbiotic Pontoniinae and, in particular, in the numerous species which gather mucus and debris from the surface of their hosts, the second legs do not seem to play an important role in feeding. Instead, they are specialized for other functions, usually for defending and identifying their own particular territory, but sometimes for additional functions such as maintaining their position on the host.

In *P. amymone* De Man the second legs are long and slender with several interdigitating teeth on the chelae. The lower border of the palm of the first legs possesses rows of anteriorly curving setae armed with setules along their forward margins. The adjacent, lower distal portion of the carpus bears a few longer but similar setae which project anteriorly. When the cheliped is flexed, this collection of setae is held dorsally up against the mouth parts. These setae are often seen to be entangled with silt and amorphous material identical to that found in the stomach. The last two joints of the endopod of the third maxilliped have prominent rows of stout setae and may also function in food gathering. The palaemonid shrimp *Leander serratus* has similar setae on the third maxilliped and although this species commonly collects large-sized food particles with its chelipeds, it does occasionally scrape up fine material directly with the third maxillipeds (Borradaile 1917). The propodi of the walking legs of *P. amymone* lack the posteriorly directed spines characteristic of free-living members of the genus.

Periclimenes spiniferus De Man is abundant both on *Pocillopora* colonies and among dead and overgrown pieces of coral and is probably one of the most common decapods on the reef flats. As in free-living species, the propodi of the walking legs possess a row of spines that may be useful in providing friction against soft substrates. As in *P. amymone* there is a tuft of comblike setae on the lower margin of chela of the first legs, which are held up against the mouth parts when the cheliped is flexed. In addition, however, the chelae themselves are modified, being flattened and enlarged, with each finger containing a row of comblike fine teeth (Borradaile 1917). The dactyl can be opened to a wide angle and the chela itself is probably used in food collection as evidenced by the presence of soft, particle-containing material on the teeth of most of the chelae examined. The fingers contact each other at their medial margins, thus forming a spoon-shaped depression on the outer surface of the chela.

The second legs of *P. spiniferus* are asymmetrical. On the chela of the larger one, the dactyl contains a peculiar pit with raised chitonous edges which opposes a similar depression on the fixed finger (Borradaile 1917). It makes a faint snapping sound when closed and thus probably has a territorial or defensive function. The fingers of the smaller chela sometimes contain a small snapping device, but more frequently each is armed with a chitonous ridge and a few proximal teeth. Both second legs are held in a fairly outstretched position (Fig. 2) and are rendered conspicuous by white circular stripes at the end of each joint and at the base of the fingers.

P. madreporae Bruce (=*P. inornatus* in Patton 1966) is translucent with a faint pinkish cast and is the smallest coral associate collected. As with *P. amymone*, the propodus of each walking leg is devoid of spines. Although its stomach contents seem to be about the same as the two preceding species, *P. madreporae* does not appear to have any special structures on the first legs for collecting or handling mucus or fine material. The terminal joint of the third maxillipeds, however, has many setae, and debris was seen entangled on them and on the setae of the second maxillipeds. The chelae of the second pair of legs have short fingers armed with a few teeth and do not appear in any way remarkable.

T. amboinensis is a shrimp of the Family Hippolytidae that is widely associated with sea anemones. It has a brown body containing large spots of gold each of which is surrounded by a thin band of light blue. The specimens from coral are smaller than those from sea anemones and differ slightly in morphology and color pattern. They may well represent a distinct species but will be considered here to be identical with those on anemones. The walking legs of *T. amboinensis* have spines on the propodus, and dactyls ending in two sharp points, and do not seem in any way modified for life on coral. The first two pairs of legs are chelate as in the Pontoniinae, but here the second legs are slender and flexible while the first are shorter and stouter. Although the stomach contents of *T. amboinensis* suggest a diet of sediment and coral mucus, there are no brushes of toothed setae on the chelae of the first or second legs. The fingers of both pairs of chelipeds, however, bear recurved setae on their distal margins which often contain debris. The chelae could thus pick up material from the surface of the colony or perhaps directly from the substrate. Individuals on sea anemones move with seeming facility from this host to the surrounding substrate[*] and it is likely that specimens on coral may do the same. If so, this could account for their rarity in night collections.

There is an additional possibility for feeding in this species. After leaving Heron Island for the last time, I had an opportunity to observe specimens of *T. amboinensis* maintained on anemones in aquaria by Mr. Hadley. Specimens were seen perched upright on the surface of the anemone, weaving their antennular flagellae about in the water. Periodically the antennules were passed down through the maxillipeds. The external antennular flagellum is enlarged and possesses long setae and is clearly being used in filter feeding. Antennular suspension feeding was not noticed in specimens of *T. amboinensis* from coral, but then it was never specifically looked for and may well occur.

Sedentary Shrimp

Harpiliopsis beaupressi (Audouin) is a member of the Pontoniinae that is much modified in association with life on coral. Compared with the agile shrimp above, this animal is very sluggish and lies with the body held closely against a coral branch. It is also somewhat softer bodied and more compressed dorsoventrally. The body is translucent with a slight brownish cast and possesses discontinuous longitudinal stripes of cream and brown, producing a form that blends in beautifully against the light brown of the coral. The walking legs are relatively shorter and stouter than those of the agile shrimps and instead of extending down beneath the body, they extend out laterally from it, wrapping around a coral branch. The propodi of the walking legs lack spines and the dactyls are stout, curved, and concave. The concavity has a row of simple setae along one side and has its inner surface covered with small serrated bumps.

As in many other members of the Pontoniinae, the dactyl of the second pair of legs is on the outer side of the chela and moves more or less horizontally (Borradaile 1917). The dactyl here is unusual in being bowed downward and having a sharp ridge of chiton on the outer lower margin. When the chelipeds are in the normal flexed position (Fig. 2), this ridge is ventral and rests directly on the coral. The probable selective advantage of the ridge is that it permits a much firmer grip on the coral than would be possible with a smooth rounded appendage. Additionally and perhaps alternatively, the ridge may serve to make the dactyl more rigid, creating a louder

[*]D. Hadley: personal communication.

snap when it is pulled against the fixed finger. The tips of the fingers are a yellow-orange color while the teeth on the inner margins are sharp and strongly chitonized. As in *P. amymone* and *P. spiniferus*, the lower margin of the chela of the first pair of legs bears a group of anteriorly curving spines with bristles along one margin. These are often tangled with soft material and silt particles, and are placed against the mouthparts when the cheliped is flexed.

Alpheus lottini Guerin (=*A. ventrosus* in Patton 1966) is a snapping shrimp of the Family Alpheidae. It is one of the best-known coral symbionts, occurring on pocilloporid corals from the Gulf of California across the Indo-Pacific to the Red Sea. As in other members of the genus, the first pair of legs are the largest and markedly asymmetrical, with the larger cheliped possessing a well-developed snapping structure and the smaller one a more normal chela. The second legs are also chelate but are much smaller and more mobile.

As Coutière (1899) has noted, *A. lottini* is well adapted to its mode of life and differs from free-living alpheid shrimp in a number of ways. It is stouter, has a smoother body, and is quite compressed laterally. The large first legs are laterally flattened and are nearly equal in size. In most species of the genus, the leg with the snapping claw is much larger than the one bearing a conventional chela. The walking legs have blunt, rather than sharp dactyls, but do not wrap around a coral branch as do those of *H. beaupressi*. Specimens seem quite sedentary, but when disturbed, they can contract their abdomen repeatedly and slither rapidly backward between the branches. The body is orange with a black dorsal stripe, while the chelae bear circular red spots that resemble coral polyps.

With regard to feeding habits, Coutière (1899) found eggs in the stomach of a specimen of *A. lottini* and implies that this species is carnivorous, while Barry (Castro 1971) found that Hawaiian specimens fed "mostly on small invertebrates and algae from the coral branches, but that coral mucus and tissue are also ingested." The stomachs which I have dissected suggest that the principal food is debris and coral mucus. Possible structures for collecting such material are the last joints of the third maxillipeds which (as in three other species of *Alpheus* examined) bear row upon row of stout setae on their inner margins and the slender, flexible, second legs which have a relatively large tangle of setae of varying lengths on the fingers of the chelae. Both sets of setae contained debris and amorphous material. A starved specimen placed on a piece of *Pocillopora* was seen to scrape the surface of the coral with one of its second legs and then clean this appendage with its mouthparts.

A. lottini is the largest of the shrimp associated with *Pocillopora* and is found between the larger branches near the base of the coral clump. A given colony usually contains only a single pair of shrimp (male and female) regardless of whether the colony is small or very large (Table 1). The members of a pair are usually seen resting close to each other. For the male-female pairs shown in Table 3, there was a significant correlation (Spearman r) between the

TABLE 3
Size and occurrence of three *Pocillopora* symbionts[*]

Colony size and number (see Table 1)	*Alpheus lottini*	*Trapezia cymodoce*	*Trapezia ferruginea* form *areolata*
Small Colonies			
309	M,F	8J,7J,4J	9M,10F
333	20M,24F		11M,13F
341	11M,18F	15F	11M,12F
342	15M,16F	8J	9M,10F
355	16M,18F		8M,10F
356	17M,17F		11M,12F
Medium Colonies			
304	18M	12F,8J,5J	8F
306	14M,11F,13F	4J,4J,3J	5J
307	15M,15F	12M,13F	
311		15M,14F,10M,10F	11M,12F,4J,4J
314		15M,15F	
315	20M,16F	18M,17F	
332	14M,15F		10M,10F
334		13M,19F	
338	28F	13M,18F	10M,13F,6J
345	2	10F,12F	13M,12F
349	16M	M,F,10M,7J,6J,6J	
354	21M,21F	16M,22F,9J,7J	
360	M,F	12M,14F,8J	8F
512	M	18M,16F,10M,11F,8J, 7J,7J,6J,5J,4J,4J	
W8	18M,20F	18F,6J	
W9	22M,14F	18M,18F,11M,12F,7J, 5J,5J,3J	11M,15F,13F
W16	22M,26F	16M,16F,12F,8J,6J	11M,12F,9F
W25	16M,17F	4J	8M
Large Colonies			
317	25M,30F	19M,20F	
329	2	12M,12F	
335	18M	17M,18F,9J,8J	
353	26M,21F	21M,23F,8J,8J,4J	
357	17M,20F	16M,18F,13M,12F, 11F,7J,5J,5J	11F
361		16M,21F,10M,13F	
513	25M	13M,19F,10F,7J, 5J,4J	
520	27M,31F		12M,14F
W2	26M,33F	15M,16F,4J	9M,10F
W15	24M,26F	7J,6J,5J,5J	14M,12F

size of the male and female components. This is what would be expected if most pairs were established early and maintained for a relatively long time.

TABLE 3 (Cont.)

Colony size and number (see Table 1)	Alpheus lottini	Trapezia cymodoce	Trapezia ferruginea form areolata
Very Large Colonies			
313	25M,29F	16M,22F,11M,16F, 14F,14F,9J, 7J,6J	10M,11F
316		18M,19F,11F	
328	M,32F	15M,19F,10F	11M,12F
331	19M,25F	16M,14F,13M,13F,10M	5J

*M=mature male; F=mature female; J=juvenile. Size=length in mm from base of rostrum to base of telson for A. lottini; carapace width in mm for the Trapezia spp.

Crabs

All of the crabs collected from living coral belong in the Xanthidae, a family with many small species that is well represented on coral reefs. In the species discussed below, the dactyl of the walking leg has a flange on its outer (posterior) surface which fits over a flange on the propodus. This apparatus, which has been described in detail by Borradaile (1902), gives the dactyl great resistance to a sideways pull, while permitting it to be held either straight, at a right angle to the propodus, or in any intermediate position. When a crab is moving over the coral, it is relatively easy to dislodge. When locked in place around a branch, however, the crab can only be removed by applying sufficient force to break the carapace or tear off a number of legs.

Chlorodiella nigra (Forskal) and *Phymodius ungulatus* (H.Milne Edward are fairly typical reef-dwelling xanthids with a carapace considerably wider than it is long. They are common free-living animals, but inhabit the living as well as the dead portions of coral colonies. As in a number of other species, the fingers of the chelipeds end in hoof-shaped tips. I believe that this is an adaptation for collecting and feeding on sediment, since I have seen several xanthids with this type of cheliped scraping up sediment and transferring it to the mouthparts. The presence of sponge tissue in the stomach of a specimen of *C. nigra* indicates that this species is probably fairly omnivorous and not restricted to a diet of sediment. When on living coral, however, it would certainly be possible for it to feed on the debris-laden mucus being shed by the coral.

Cymo andreossyi (Audouin) is much more sluggish than the above two crabs and appears more adapted to a coral-dwelling existance. It is found on pocilloporid corals and occasionally on dead coral but not on the branching species of *Acropora*. It is a cream-colored crab with a rather flat and subcircular carapace and with abundant setae on the walking legs giving it a shaggy appearance. The chelipeds are markedly asymmetrical. One is exceptionally large and heavy, with stout fingers containing a few heavy rounded teeth and ending in blunt, hoof-shaped tips. The other cheliped is considerably smaller and has thin, slender fingers that terminate in hoof-shaped tips, and have opposing lateral flanges. The large chela of *C. andreossyi* seems well suited for breaking off pieces of coral. No zooxanthellae or coral tissue remains were seen in the stomachs, however, and thus there is no indication that they do break off coral. According to the stomach contents, the most likely food would seem to be sediment and coral mucus collected with the smaller cheliped.

C. andreossyi sometimes occurs in pairs, but frequently does not. Usually only one to three individuals are found per colony although one colony contained twelve small specimens (Table 1). This species is heavily infected with a rhizocephalan parasite, *Loxothylacus variabilis*. Of twenty specimens with a carapace width of over 10 mm, twelve had an external parasite and were thus removed from the breeding population (Patton 1966).

Crabs of the genus *Trapezia* have a smooth, shiny, and in most species, brightly colored carapace that is not much broader than it is long. They are typically found on corals of the Family Pocilloporidae and extend in range from the east coast of Africa across the Indo-Pacific to the west coast of North America. They are most commonly found in the central and basal portions of the coral colony. They can hold onto the coral very firmly with their walking legs, but can also move around in the branches with considerable speed and agility. The dactyls of the walking legs are broad and blunt and bear rows of blunt spines on their inner surface. Knudsen (1967) observed these spines to be used in the induction and transportation to the mouthparts of host mucus secretions. The chelipeds are relatively long, with thin sharply pointed fingers and are extended outward at any intruder. If the crab is given the opportunity, it will deliver a rapid and tenacious nip to the finger of the collector.

The species of *Trapezia* show a tendency toward pair formation accompanied by territoriality with the result that, as in *A. lottini*, a colony is often occupied by only one mature male and one mature female of a given species. *Trapezia* shows the phenomenon of color forms in which morphologically similar specimens possess markedly different color patterns. These color forms are actually distinct species, as both members of a pair always belong to the same form.

Trapezia cymodoce (Herbst) was the largest and most common species of the genus found at Heron Island. It has a blue-gray carapace and yellow-orange legs and ventral surface. The fingers of the chelipeds are brown and when opened in threat posture reveal a conspicuous white spot between them. Each inhabited *Pocillopora* colony usually contained a single largest pair although smaller adults and juveniles often co-occurred (Table 3).

   T. ferruginea form *areolata* Dana is a distinctly smaller species in which the carapace and dorsal surface of the chelipeds are covered with a prominent network of thin red lines. Pairing is more strict than in *T. cymodoce,* with more than half the colonies collected containing only a single pair and with juveniles being less common (Table 3).

   In view of the smaller size of *T. ferruginea* form *areolata,* it is most interesting to note that a pair of adults occurred on all the small colonies of *Pocillopora* shown in Table 1, but on only 34% of the larger colonies. In the larger *T. cymodoce,* however, no pairs were present on the small colonies, while 66% of the larger colonies contained one or more pairs. *T. ferruginea* was much less common than the preceding two species.

   Although not well camouflaged when away from living coral, the *Trapezia* species seem structurally well equipped to move between colonies. The extent to which they do so is an interesting subject for future research. As in the case of *A. lottini,* the fairly close correlation between the sizes of the male and female components of a pair (Table 3) suggests that the pairs are formed early and are relatively long lasting, rather than being repeatedly reformed between different individuals.

   With regard to feeding habits of the *Trapezia* species, Crane (1937) found worms in the stomachs of two species collected from the Gulf of California. Gerlach (1960) places *Trapezia* and the other crabs and shrimp which regularly inhabit coral branches in the category of detritus feeders and small predators but does not give further details. Knudsen (1967) made direct observations on mucus feeding in *T. ferruginea* and *T. ferruginea* form *areolata* at Eniwetok. When starved specimens were placed on *Pocillopora,* they were seen to scratch the coral tissue with the dactyls of the walking legs. Coral mucus adhering to the spines and setae of the dactyls was cleaned off by the mouthparts. Occasionally the tips of the chelae also transferred material from the coral surface to the mouthparts. My own studies of stomach contents indicated that the three most common *Trapezia* species feed on coral mucus and on sediment settling on the host. Laboratory observations with *T. cymodoce* indicated that it is not a complete specialist. Pieces of shrimp tissue dropped on the coral were caught basket fashion with the walking legs and transferred to the mouthparts. Several possible scraping movements of the dactyls against the coral tissue were seen but not with the regularity observed by Knudsen (1967).

Fish

   Paragobiodon echinocephalus Ruppell is a small fish associated with pocilloporid corals. Of the specimens taken from *Pocillopora damicornis,* some individuals were yellow with black fins and a faintly orange head, while the majority were black with an orange-red head. The pectoral fins are united in the ventral midline, forming an adhesive disc. This device is efficient enough to allow individuals to remain attached for long periods to the side walls of glass aquaria and it doubtless aids the animal greatly in maintaining its position within the coral colony. The existence of a single male-female pair per colony is more common than indicated in Table 1 and represents about half of all the occurrences of this species.

Hiatt and Strasburg (1960) found that of three *Paragobiodon* with food in their stomachs, two contained isopods and one shrimp fragments; they concluded that the species is highly carnivorous. This view is supported by the fact that specimens in the laboratory can be seen catching *Artemia* nauplii. The stomachs which I examined, however, did not yield any remains of Crustacea or other animals. Instead, yellow cells 5-7 µ in diameter that must surely represent coral zooxantheallae were common and three specimens contained cross sectional views of coral tentacles with intracellular zooxanthellae clearly visible. The stomach of a specimen that had been starved for three days and then placed on a piece of *Pocillopora* contained several concentrations of zooxanthellae that probably represented pieces of coral tissue. While some of the zooxanthellae found in *Paragobiodon* stomachs may have been expelled from coral tissues, it is apparent that this species sometimes ingests intact tissue and is thus partially parasitic upon its host. The small size of the fish, combined with the regenerative powers of the host and the fact that other types of food may be taken, evidently permits this relationship to be a stable one.

DISCUSSION

The mobile macroscopic animals associated with *Pocillopora* show varying degrees of adaptation to their host. Of the most common species, *Periclimenes spiniferus*, *C. nigra*, and *Phymodius ungulatus* are basically free-living species and are commonly found away from living coral as well as on it. They show no particular structural modifications for life on living coral and presumably move freely onto and away from it. A second group includes *P.amymone*, *P. madreporae*, *T. amboinensis*, and *C. andreossyi*. They are most easily found on living coral and are apparently most abundant there. They appear relatively unmodified but do show slight morphological differences from typical free-living species. The last three show a degree of host specificity, being more common on pocilloporid corals than on the branching species of *Acropora* (Patton 1966).

The third group consists of *H. beaupressi*, *A. lottini*, the *Trapezia* species, and the fish, *Paragobiodon echinocephalus*. All are normally found only on pocilloporid corals, and in the first two cases at least, are restricted to certain genera within the family. At Heron Island, *H. beaupressi* inhabits *Pocillopora* but not *Seriatopora* or *Stylophora*, while *A. lottini* inhabits the first two genera but not the last (Patton 1966). They show many structural adaptations

to their host and adults appear not to leave a given coral colony
once they are established upon it. Except for the first species
they show a tendency (best developed in *A. lottini*) to occur in single male-female pairs per colony. In such a case a given pair presumably maintains its colony as a territory and defends it against
conspecific individuals. *Pocillopora* colonies are quite open and
presumably fairly easy to patrol. They harbor only a single pair of
A. lottini, while the related *Seriatopora hystrix*, which forms a
thicket of interlacing branches, was on several occasions found to
harbor two pairs.

The sediment between corals on the Heron Island reef flat is
surprisingly fine, and as indicated by the filling up of cans placed
on the reef, considerable amounts of this material are continually
being suspended and redeposited. Corals can cope with this rain of
sediment by means of ciliary currents supplemented by mucus secretions which together carry particles off the surface of the colony
(Marshall and Orr 1931; Yonge 1930). The crustaceans on *Pocillopora*
ingest this material and may even benefit their host slightly by thus
accelerating the removal of sediment. Although Marshall and Orr
(1931) found no more than 6% organic matter in the sediment which
collected in jars at Low Isles, the large amounts of mineral matter
in the stomachs of several species suggest that considerable amounts
of sediment may be processed. Suspended particulate organic matter
(organic aggregates) is common in the water around coral reefs and
is fed upon directly by various fish (Johannes 1967) and by the coral
crab *Domecia acanthophora* (Patton 1967). Some of this material must
settle onto the coral and it probably accounts for part of the soft
material found in stomachs of commensal decapods. That material settling on the coral colony has nutritive value is indicated by the
fact that several reef corals (including *Pocillopora*) have ciliary
currents which can convey material from the surface of the colony
into the polyps (Yonge 1930). Johannes (1967) believes that coral
mucus is the source of organic aggregates occurring behind the reef
at Eniwetok and estimates that corals may shed mucus in an amount
equivalent to 2% of the total production of the reef community. If
Pocillopora is indeed exporting large amounts of organic material,
then its crustacean associates are in the fortunate position of being
located at the source of this flow.

The relative importance of sediment and coral mucus in supplying
the energy requirements of the various crustaceans found on *Pocillopora* is, of course, unknown. Coarse mineral particles predominate
in the stomachs of *Periclimenes spiniferus, C. nigra,* and *C. andreossyi*, indicating a diet of sediment. In crustaceans of group three,
however, soft material dominated the stomachs examined and thus coral
mucus might be the major energy source.

The associates of *Pocillopora* represent stages in the evolution
of symbiosis. The group one species are free-living animals whose
deposit-feeding habits make it possible for them to feed while on
living coral. Group two contains species which may or may not live
in sufficiently close association with *Pocillopora* to be regarded as

symbionts. The species of group three are clearly obligatory symbionts of living coral and show many modifications associated with
their habitat. The crustacean members of this group may well have
evolved a dependence on the mucus produced by their host. Since they
do not feed directly on living host tissue, they still merit their
usual designation as commensals. By contrast, the fish, *Paragobiodon
echinocephalus*, while still quite capable of catching small organisms
in the manner typical of other gobies, appears to rely on coral tissue as its major food source and is thus parasitic.

Table 4 summarizes the occurrence of coral associates on different sizes of *Pocillopora* colonies. Assuming that the coral grows
in a hemispherical shape and encloses a volume consisting half of
branches and half of spaces between branches, then the figures given
represent conservative estimates of the ratio of increase of available space as the colony grows larger. It can be seen that the number of individuals per colony increases about as would be expected
in moving from the small to medium-sized colonies, but that the rate
of increase drops sharply thereafter. This indicates that there must
substantial amounts of unutilized space within the larger colonies.
The number of species per colony is about the same in the three
larger-sized colonies. The increase in individuals that does occur
is due largely to increasing numbers of *P. spiniferus*, *P. madreporae*,
H. beaupressi, and *C. nigra*. *A. lottini* and *T. ferruginea* form *areolata*, however, must have to patrol a larger area on larger colonies,
but still maintain a single male-female pair per colony. The remaining species increase their numbers not at all or relatively little.
The combination of these factors results in a less even distribution
of individuals into species as colony size increases and in a declin-
ing value of (J).

TABLE 4
Effect of colony size on *Pocillopora* associates[*]

Colony size	Small	Medium	Large	Very Large
Number of colonies	6	18	10	4
Approximate colony diameter in centimeters	12.0	20.0	30.0	40.0
Ratio of space available	1	3	8	20
Mean number of individuals per colony	7.7	21.0	33.0	65.0
Mean number of species per colony	4.5	7.5	7.5	8.3
Mean diversity (H)	.41	.60	.61	.65
Mean evenness (J)	.99	.90	.85	.83

[*]Data from Table 1.

ACKNOWLEDGMENTS

I am most grateful to the Great Barrier Reef Committee for permission to use the Heron Island Research Station and to the members of the Department of Zoology of the University of Queensland for many kindnesses. My first trip to Australia was supported by a Fulbright Scholarship from the United States Educational Foundation, Canberra, and the second aided by funds from N.S.F. Grant GY7673. Figure 1 was drawn by Mr. R. Greet.

LITERATURE CITED

Borradaile, L. A. 1902. Marine crustaceans. III. The Xanthidae and some other crabs. In Fauna and geography of the Maldive and Laccadive Archipelagos 1:237–71, ed. J. S. Gardiner. Cambridge: University Press.

_____. 1917. On the Pontoniinae. The Percy Sladen Trust Expedition to the Indian Ocean in 1905, under the leadership of Mr. J. Stanley Gardiner. Trans. Linn. Soc. Lond. Zool., ser. 2, 17: 397–412.

Bruce, A. J. 1971. Notes on some Indo-Pacific Pontoniinae. XVII. Eupontonia noctalbata gen. Nov., Sp. Nov., A new pontoniid shrimp from Mahè, The Seychelle Islands. Crustaceana 20:225–36.

Castro, P. 1971. The natantian shrimps (Crustacea, Decapoda) associated with invertebrates in Hawaii. Pacif. Sci. 25:395–403.

Coutière, H. 1899. Les Alpheidae, morphologie externe et interne, formes larvaires, bionomie. Ann. Sci. Nat. Zool., ser. 8, 9: 1–559.

Crane, J. 1937. The Templeton Crocker Expedition. III. Brachygnathous crabs from the Gulf of California and the west coast of Lower California. Zoologica 22:47–78.

Gerlach, S. 1960. Über das tropische Korallenriff als Lebensraum. Zool. Anz. Suppl. 23:356–63.

Goreau, T. F. 1956. Histochemistry of mucopolysaccharide-like structures and alkaline phosphatase in Madreporaria. Nature 177:1029–30.

Hiatt, R. W., and Strasburg, D. W. 1960. Ecological relations of the fish fauna on the coral reefs of the Marshall Islands. Ecol. Monographs 30:65–127.

Johannes, R. E. 1967. Ecology of organic aggregates in the vicinity of a coral reef. Limnol. Oceanog. 12:189–95.

Johnson, D. S., and Liang, M. 1966. On the biology of the watchman prawn, Anchistus custos (Crustacea; Decapoda; Palaemonidae), an Indo-West Pacific commensal of the bivalve Pinna. J. Zool. Lond. 150:433–55.

Knudsen, J. W. 1967. Trapezia and Tetralia (Decapoda, Brachyura, Xanthidae) as obligate ectoparasites of pocilloporid and acroporid corals. Pacif. Sci. 21:50–57.

Marshall, S. M., and Orr, A. P. 1931. Sedimentation on Low Isles reef and its relation to coral growth. Sci. Rept. Gt. Barr. Reef Exped. 1:93-133.

Patton, W. K. 1966. Decapod Crustacea commensal with Queensland branching corals. Crustaceana 10:271-95.

_____. 1967. Studies on *Domecia acanthophora*, a commensal crab from Puerto Rico, with particular reference to modifications of the coral host and feeding habits. Biol. Bull. 132:56-67.

Pielou, E. C. 1966. The measurement of diversity in different types of biological collections. J. Theoret. Biol. 13:131-44.

Yonge, C. M. 1930. Studies on the physiology of corals: I. Feeding mechanisms and food. Sci. Rept. Gt. Barr. Reef Exped. 1:13-57.

Symbiotic marine algae: taxonomy and biological fitness

Dennis L. Taylor
Rosenstiel School of Marine and Atmospheric Science
University of Miami
Miami, Florida

Algal-invertebrate associations present a unique opportunity to explore the fundamentals of multicellularity in comparably simple and relatively well-defined systems of cellular interaction (Taylor 1973). In theory, they are best regarded as complete functional units, while in practice their greatest virtue lies in the ease with which the participants can be recognized, separated, and examined in vitro. Exploitation of such experimentally useful properties has contributed substantially to our knowledge of these associations during the last decade.

Particular progress has been made in the area of nutrition and nutrient exchange (Smith, Muscatine, and Lewis 1969; Taylor 1971a; Trench 1971a, b, c), a subject that is central to very many problems in symbiosis. Because the animal is generally the major partner (by virtue of size), these studies have tended to favor the nutritional requirements of the host-recipient. Historically, algal symbionts have been viewed simply as donors, and their biology has been generally neglected. Few workers feel happy if the identity of the host is unknown to them. Most are content to refer to marine algal symbionts as "zooxanthellae," a term that lost all of its taxonomic significance before the end of the nineteenth century. This is more than merely unfortunate. These algae serve as one of the primary vehicles for energy input in symbiosis (e.g., Goreau, Goreau, and Yonge 1971). Knowledge of their taxonomy, distribution, biochemistry, and physiology would provide a basis for assessing the quantity and quality of this input, and aid in understanding how these are regulated in time and space. Such information places the basic question of host nutrition on a much firmer footing, making it more amenable to investigation. The same data could well elucidate the evolutionary history of the host, or provide a point of attack for solving the species problem in a difficult group such as the hermatypic corals (e.g., through the examination of light-dependent growth forms, and the quality of symbiont photosynthate with depth). Recently, these data have been used to advantage as a means of exploring host specificities and nutritional dependence in the flatworms *Convoluta roscoffensis* and *Amphiscolops langerhansi* (Provasoli, McLaughlin, and Droop 1968; Taylor 1971b; Nozawa, Taylor, and Provasoli 1972). Clearly, the possibilities for unveiling significant aspects of a vast majority of associations via this avenue of attack are both numerous and varied.

Essentially, the solution to the present experimental bottleneck lies first with the isolation, cultivation, and identification of algal symbionts, and second with a detailed study of their biology in culture and in vivo. The foundations for this have been in progress for nearly thirty years. Kawaguti (1944) was the first to isolate and observe the motile stages of *Gymnodinium (=Symbiodinium) microadriaticum* (Freudenthal), possibly the most ubiquitous of marine algal symbionts (see below, p. 242). He was, however, unable to bring these into culture. Subsequently, very little success was achieved. The design of artificial seawater media for the cultivation of marine algae (Provasoli, McLaughlin, and Droop 1957) eventually provided the necessary expertise, and shortly thereafter *G. microadriaticum*, isolated principally from *Cassiopeia frondosa* and *Condylactis gigantea*, was established in axenic culture (McLaughlin and Zahl 1957, 1959, 1962a, b; Zahl and McLaughlin 1957, 1959). Freudenthal (1962) provided the first taxonomic description of a symbiont in culture, and McLaughlin and Zahl (1966) utilized the same material in physiological and cultural studies. In recent years, the range of marine symbionts in culture has increased modestly (Parke and Manton 1967; Norris 1967; Taylor 1968, 1969a, b, c, 1971b, c). The number of different species is few (see below) and, as Droop (1963) has noted, the majority of these are somewhat stereotyped in contrast to the almost incredible diversity of hosts. At the present time, our knowledge of these organisms is sufficient to enable us to classify them with a reasonable degree of certainty, and to draw some general conclusions about the nature of their distribution among various invertebrate phyla. Recent studies of symbionts in culture suggest that it may even be possible to pose a tentative answer to the question, "what are the properties of good algal symbionts?"

TAXONOMY

Some appreciation of the range and variation among marine algal symbionts can be gained from an examination of those figured in an early paper (Brandt 1883) (Fig. 1). Representatives from at least

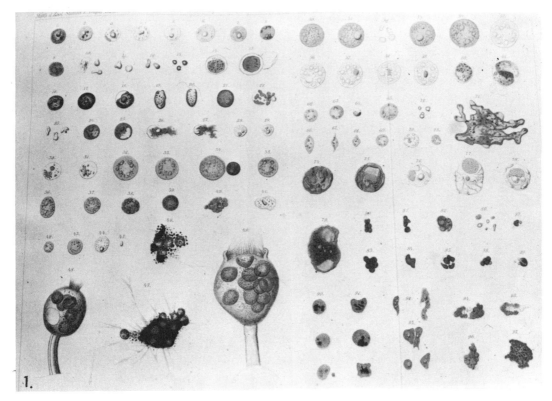

Figure 1. Algal symbionts recorded by Brandt (1883). "Zooxanthella"
types *Gymnodinium microadriaticum* (72-74), *Amphidinium
chattonii* (32-35), Cryptophyceae (19-20), ameboid **Chrysophyceae** (71, 94-97), Chlorophyceae (90-93).

four, possibly five (if one considers the recent work of Ax and
Apelt 1965; Apelt 1969), algal classes appear to be present (i.e.,
Bacillariophyceae, Chlorophyceae, Chrysophyceae, Cryptophyceae, and
Dinophyceae). Excluding the duplication of stages in individual life
histories, it is possible to recognize species that have been described in recent years (see explanation, Fig. 1). On the other hand,
several are unknown to us. Their existence today has not been confirmed in an examination of the host species described by Brandt
(1883) from the Bay of Naples (Taylor, unpublished). With the exception of cryptomonad like motile cells (Fig. 1, nos. 19 and 20),
ameboid, chrysomonad like stages (Fig. 1, Nos. 71 and 94-97) and
obvious chloromonad like symbionts (Fig. 1, nos. 90-93), Brandt's
figures are almost exclusively of yellow-brown, coccoid nonmotile

cells-the classic "zooxanthella". These are by far the most ubiquitous of all the known marine algal symbionts, and constitute the most
important major grouping. Their taxonomic placement among the dinoflagellates has been known for a considerable period (Klebs 1884;
Chatton 1923; Kawaguti 1944; Pringsheim 1955; McLaughlin and Zahl
1957), although they have been frequently confused with the crypto-
monads (Fritsch 1935, 1952; Caullery 1952). Representatives of the
remaining algal classes tend to be a very minor component of the symbiotic microflora. Acoelus Turbellaria apparently harbor the only
known symbiotic marine Chlorophyceae (Parke and Manton 1967) as well
as Bacillariophyceae (Ax and Apelt 1965; Apelt 1969) and Dinophyceae
(Taylor 1971b). Symbiotic representatives of the Cryptophyceae and
Chrysophyceae are extremely rare, poorly studied, and of doubtful
validity. Species of marine blue-green algae (Cyanophyceae) may
also enter into symbiotic associations (Lebour 1930; Sará and Liaci
1964; Sará 1971; Taylor, unpublished; Droop 1963), but these are
poorly studied and have not been included in the present communication.

Major Symbionts

For present purposes, the algae considered as "major symbionts"
include only those species that are readily assignable to the coccoid, yellow-brown "zooxanthellae." They are by far the most frequently encountered marine algal symbionts, and occur most widely
among protozoa and coelenterates.

Gymnodinium microadriaticum (Freudenthal)

G. microadriaticum is certainly the best known of all the marine
symbionts. The first to be isolated in axenic culture (McLaughlin
and Zahl 1957, 1959, 1962a, b; Zahl and McLaughlin 1957, 1959), it
was originally described as *Symbiodinium microadriaticum* (Freudenthal
1962). Recent work with cultures of *S. microadriaticum* and fine
structural evidence shows that its affinities with *Gymnodinium* are
such that its placement in a separate genus can hardly be justified
(Taylor 1971b, c). As a general rule, the symbiotic habit should
not be viewed as a primary taxonomic character (Taylor 1971c).

In culture, *G. microadriaticum* gives rise to motile gymnodinioid
cells. The encysted phase predominates, and is found to be similar
in morphology to the symbiont in vivo. The life history in culture
has been described by Freudenthal (1962), and consists of a simple
alternation between motile and encysted stages without apparent cyc-
lic patterns. The production of gametes and the possibility of sexual reproduction hinted at by Freudenthal (1962) have been confirmed in our laboratory. Motile stages have never been recorded in
nature and are presumed to be extremely rare, although they could
serve as a mechanism of dispersal. Reports by Kinsey (1970) that
planulae of the gorgonians *Briarium asbestinum* and *Muriceopsis fla-
vida* lack symbionts deserve close scrutiny since they may lead to a
better understanding of the production of motile *G. microadriaticum*

in nature.

    The principal morphological features of the cell may be seen in Figure 2 (see also Taylor 1968, 1969<u>a</u>; Kevin et al, 1969). Its fine

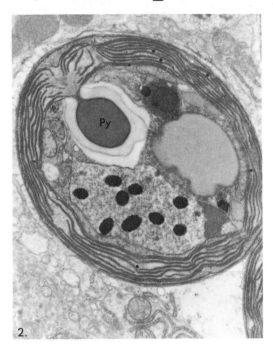

Figure 2. Fine structure of *Gymnodinium microadriaticum* showing the pyrenoid (Py) and its attachment to the inner face of the chloroplast.

structure is typical of the dinoflagellates. The key taxonomic character is the possession of a single, stalked pyrenoid, projecting from the inner surface of the chloroplast and encased in a cap or sheath of starch. This feature is clearly seen in the optical microscope, and in conjunction with the form of the motile stage, it is a reliable means of identifying the species. Numbers 74 and 75 in Brandt's illustration (Fig. 1) are representative of *G. microadriaticum*. The alga's distribution is given below (p. 249).

Amphidinium chattonii (Hovasse)

    Among pelagic hosts, the dinoflagellate genus *Amphidinium* constitutes the major symbiont group (Taylor 1969<u>b</u>). *A. chattonii* (formerly *Endodinium chattonii*) is the only completely described species, although at least one other is suspected to exist (see below, p. 244). Early studies were based solely on the morphology of the alga <u>in situ</u> (Hovasse 1923), and until recently, the existence of a motile stage was unknown (Taylor 1971<u>c</u>). Studies of the alga in culture demonstrate an amphidinium motile stage having the klebsiicarterae morphology (Taylor 1971<u>d</u>). In typical cultures, this constitutes less than 10% of the total cells, the remainder being spherical encysted forms. The life history of the alga is similar to that noted for *G. microadriaticum*. Transmission between host generations is cellular during asexual reproduction of the host or via the egg in sexually reproducing hosts (e.g. *Velella velella*).

Figure 3 illustrates the most important taxonomic featues. Like *G. microadriaticum*, the pyrenoids are the key to distinguishing the species. In all specimens examined, these are usually four in number, and suspended from the interior of the chloroplast by multiple stalks. A cap of starch surrounds the entire structure, and the central matrix is penetrated by chloroplast lamellae. Since both the position and number of pyrenoids are easily seen in the optical

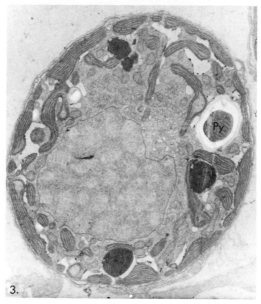

Figure 3. Fine structure of *Amphidinium chattonii* illustrating the characteristic structure of one of its pyrenoids (Py). Note the presence of plastid lamellae within the pyrenoid matrix.

microscope, the species is readily recognized in the field. A second character which makes this species of special interest to the phycologist is the possession of plastid lamellae formed from only two thylakoids instead of the normal three found in other dinoflagellates (Dodge 1968). Numbers 32 through 35 in Brandt's illustration (Fig. 1) are representative of this species.

Amphidinium sp.

A third major symbiont has been isolated from the radiolarians *Colozoum inerme* and *Sphaerozoum punctatum*; it appears to occur commonly among planktonic protozoa (cf. Doyle and Doyle 1940; Freudenthal, Lee, and Kossoy 1964; Lee et al. 1965). Like *A chattonii*, it exists primarily as a coccoid cyst, but under suitable conditions it will give rise to a motile amphidinium stage. Preliminary studies indicate that it is very similar to *A. chattonii*, differing only in the possession of a single, large pyrenoid (Fig. 4).

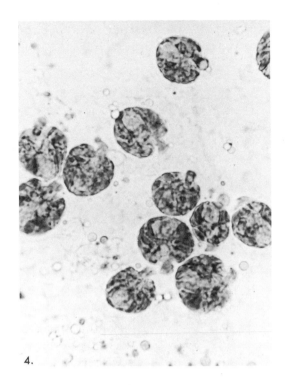

4.

Figure 4. Optical micrograph of *Amphidinium* sp. isolated from *Colozoum inerme* collected in the Bay of Naples, 1968.

Minor Symbionts

The acoelus Turbellaria appear to harbor a number of different symbiont genera, which vary considerably in the degree of cellular integration achieved within the host tissues. Although the occurrence of these animals is comparatively restricted in nature, they are easily maintained in culture, and their variation in symbiotic expression makes them valuable laboratory tools. For the present purpose, their algal symbionts have been classed as "minor" in order to distinguish them from the more widely distributed species noted above.

Amphidinium klebsii Kofoid et Swezy

The symbiont of the flatworm *Amphiscolops langerhansi* has recently been isolated in axenic culture and identified as the dinoflagellate *A. klebsii* (Taylor 1971b), formerly known only as a freeliving species. As shown in Figure 5, the symbiont is unusual in that it is completely lacking in any structural adaptations for the symbiosis. Examination of its fine structure in situ shows that even its two flagella are retained inside the host (Fig. 6). Taxonomically speaking, the alga is of interest since it reinforces the view that morphology, not the symbiotic habit, is the primary criterion for classifying these organisms. Important characters for

its identification in the field concern the structure of the pyrenoid, and its relationship with the radial chloroplast arrangement
(Taylor 1971<u>d</u>).

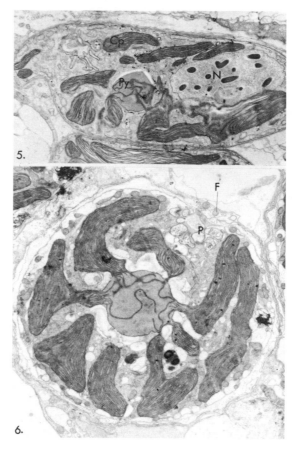

Figure 5. Fine structure of *Amphidinium klebsii* <u>in situ</u>. Longitudinal section showing typical alignment and cell shape
within the host. Chloroplast (Cp), pyrenoid (Py), nucleus
(N).

Figure 6. Transverse section of *Amphidinium klebsii* <u>in situ</u> showing
the longitudinal flagellum (F) and its associated pusule
(P).

Platymonas convolutae Parke et Manton

The algal symbiont of *Convoluta roscoffensis* has been known for
a considerable period of time (Geddes 1880; Gamble and Keeble 1904;
Keeble and Gamble 1907; Keeble 1910). It was originally identified
with the genus Carteria (Keeble and Gamble 1907); however, Parke and
Manton (1967) have presented a detailed study of its ultrastructure
both <u>in situ</u> and in culture, and place it as a new species of *Platymonas*, *P. convolutae*. Droop (1963) also recognized its affinities
with the Prasinophyceae (Christensen 1965) at an earlier date. The
association formed with the host is a highly integrated unit which,

by the strictest of definitions, is wholly autotrophic in character. Recent studies employing this alga and related species of free-living Prasinophyceae have explored the problems of host specificity (Provasoli, Yamasu, and Manton 1968), and comparative photosynthetic efficiency (Nozawa, Taylor, and Provasoli 1972), using techniques of cross infection and symbiont competition.

As noted previously, green algal symbionts are rare in the marine environment. Unpublished observations on the green symbionts found in the Pacific coast anemones *Anthopleura elegantissima* and *A. xanthogrammica* suggest that they may also be species of *Platymonas*, possibly *P. convolutae*. It seems likely that this alga is more widespread than originally suspected. *P. convolutae* has also been isolated from the flatworms *Convoluta psammophyla**, *C. macnaei*, and *C. sutcliffi* (Taylor, unpublished).

Problematic Symbionts

Symbionts considered as being problematic are those which have received tentative treatment in the literature, and still lack complete taxonomic details. As a group, they are extremely unusual. If valid, they are probably exceptionally rare.

Licmophora sp.

Ax and Apelt (1965) describe a species of the diatom genus *Licmophora* symbiotic with specimens of the flatworm *Convoluta convoluta* (see also Apelt 1969). The alga has been isolated in culture. However, there are no taxonomic details at present which would facilitate its complete identification. Attempts to reinfect alga-free hosts with cultured cells have not been successful, a fact that raises some doubt as to whether *Licmophora* is the true endosymbiont of *Convoluta convoluta*.

Cryptomonas sp.

The data in support of the existence of symbiotic Cryptophyceae are not convincing. Nevertheless, these organisms have been frequently reported in associations with marine invertebrates such as *Cryptomonas schaudinni* (Winter 1907). Also Brandt (1883), Fritsch (1935, 1952); and Caullery (1952) (see Fig. 1, nos. 19 and 20) and their cause deserves some attention here. As a general rule, species of foraminifera are implicated as hosts (Brandt 1883; Winter 1907; Lee and Zucker 1969). These organisms are widespread, and much of the confusion may be due to their ability to establish multiple symbioses with a variety of micro-algae (Taylor, unpublished data). Close attention to this aspect in the field and in laboratory cultures may eventually provide positive data on symbiotic Cryptophyceae. Until that time, their existence is in doubt.

---

*M. Parke: personal communication.

DISTRIBUTION

The distribution of known algal symbionts among the major invertebrate phyla is summarized in Table 1. Since most of the records are taken from personal observations, the list is somewhat restricted. Nevertheless, it does serve to demonstrate the potential of various symbionts and their requirements regarding host type. In this respect, it is interesting to note that the two major symbionts, *G. microadriaticum* and *A. chattonii*, exhibit distinct preferences with regard to thier host's life style. *G. microadriaticum* is the most widely recorded species and is specifically associated with benthic hosts. *A. chattonii* is specifically associated with pelagic hosts, as is the closely allied *Amphidinium* sp.

SYMBIONT FITNESS

One of the most striking features of all algal symbionts in culture is their similarity to known free-living species. Despite this, experimental studies involving cross infections and situations of symbiont competition clearly indicate definite host preferences and subtle degrees of nutritional interdependence (Provasoli, Yamasu, and Manton 1968; Taylor 1971b; Nozawa, Taylor, and Provasoli 1972). Studies of algal symbionts in culture may help to resolve these apparent contradictions, and point to specific adaptations that make some algae more suitable for symbiosis than others. Generally speaking, these adaptations fall into two broad categories, structural and nutritional.

Structural Adaptations

The success of dinoflagellate symbionts in the marine environment may be partially due to selection of these forms from available free-living species, based on their rigid, multiple-layered thecal structure (Dodge and Crawford 1970), and the ease with which they can form resistant cysts. Both of these qualities are characteristic of the Dinophyceae as a whole, and may serve to protect the symbiont from possible digestion within host tissues and during periods of transmission between host generations. When these algae are introduced into asymbiotic hosts, and a symbiosis is established, no concurrent structural changes take place (Taylor 1971b). The structural adaptations of dinoflagellate symbionts are a property of the class as a whole, not a character that had evolved in response to the symbiosis. Their value to the alga, and the associations which it forms with invertebrate hosts, must be weighed against the disadvantages which such rigid structures present to the rapid movement of nutrients and waste materials in the symbiosis (Taylor 1969 c).

255

TABLE 1
Host distributions of principal algal symbionts

DINOPHYCEAE

A. *Gymnodinium microadriaticum*

Protozoa:Ciliata
    Euplotes patella
    Urceolaria patella
    Trichodina sp.
Coelenterata:Hydrozoa
    Halecium beani
    Sertularella polyzonias
    Sertularella operculata
    Aglaophenia helleri
    Aglaophenia pluma
    Myrionema sp.
Coelenterata:Scyphozoa
    Cassiopeia frondosa
Coelenterata:Actinozoa (Actinaria)
    Aiptasia couchii
    Anemonia sulcata
    Cereus pedunculatus
    Anthopleura elegantissima
    Anthopleura xanthogrammica
    Condylactis auriantica
    Condylactis gigantea
    Heliactis bellis
Coelenterata:Actinozoa (Alcyonaria)
    Briarium asbestinum
    Plexaura homomalla
    Pseudoplexaura porosa
    Eunicea mammosa
    Eunicea tourneforti
    Muriceopsis flavida
    Plexaurella pumila
    Muricea muricata
    Pseudopterogorgia bipinnata
    Gorgonia flabellum
Coelenterata:Actinozoa (Madreporaria)
    Stephanocoenia michelinii
    Stephanocoenia intersepta
    Acropora palmata
    Acropora prolifera
    Acropora cervicornis
    Agaricia agaricites
    Agaricia tenuifolia
    Agaricia undata
    Agaricia lamarckii
    Agaricia grahamae

Agaricia fragilis
Helioseras cucullata
Siderastrea siderea
Siderastrea radians
Porites astreoides
Porites porites
Porites divaricata
Porites furcata
Favia fragum
Diploria clivosa
Diploria labyrinthiformis
Diploria strigosa
Manicina aureolata
Cladocora arbuscula
Montastrea annularis
Montastrea cavernosa
Solenastrea hyades
Oculina diffusa
Meandrina meandrites
Dichocoenia stokesii
Dichocoenia stellaris
Dendrogyra cylindrus
Mussa angulosa
Scolymia lacera
Isophyllia sinuosa
Isophyllia multiflora
Isophyllastrea rigida
Mycetophyllia lamarckiana
Mycetophyllia ferox
Eusmilia fastigiata fastigiata
Eusmilia fastigiata flabellata
Fungia danae
Galaxea fascicularis
Pocillopora damicornis
Mollusca:Lamellibranchia
    Tridacna crocea
    Tridacna derasa
    Tridacna gigas
    Tridacna maxima
    Tridacna squamosa
    Hippopus hippopus
    Corculum cardissa
Mollusca:Nudibranchia
    Aeolidiella glauca

TABLE 1 (Con't.)

DINOPHYCEA	

B. *Amphidinium chattonii*
  Coelenterata:Hydrozoa
    Velella velella
    Porpita porpita
C. *Amphidinium klebsii*
  Platyhelminthes:Turbellaria
    Amphiscolops langerhansi
	D. *Amphidinium* sp.
	  Protozoa:Sarcodina
	    Collozoum inerme
	    Globigerinoides ruber
	    Orbitolites sp.
	    Peneroplis sp.
	    Sphaerozoum punctatum

PRASINOPHYCEAE	

A. *Platymonas convolutae*
  Platyhelminthes:Turbellaria
    Convoluta roscoffensis
    Convoluta psammophyla
    Convoluta macnaei
    Convoluta sutcliffi
	B. *Platymonas* sp.
	  Coelenterata:Actinozoa (Actinaria)
	    Anthopleura elegantissima
	    Anthopleura xanthogrammica

The green algal symbiont, *P. convolutae*, is unique in its contrast to the more common dinoflagellate types. Its physical relationship with *C. roscoffensis* has clearly evolved in response to the symbiosis (Droop 1963), and the general morphology of the alga in situ bears little resemblance to its free-living stages (Parke and Manton 1967; Oschman 1966). As an autotrophic unit, the success of the association rests heavily on the speed with which nutrients and waste materials can be translocated between host and alga. *P. convolutae* satisfies these demands by assuming the form of a naked protoplast. Experiments with other free-living Prasinophyceae show that they cannot achieve the same intimacy of contact with the host (Provasoli, McLaughlin, and Droop 1968). This may be an important factor in the success of *Convoluta* in competitive situations such as those introduced experimentally or in nature where reinfection of host larvae must take place from mixed algal populations.

Nutritional Adaptations

Studies of *G. microadriaticum* in culture show that the alga is able to utilize a number of different organic and inorganic sources of nitrate and phosphate (Droop 1963; McLaughton and Zahl 1966). Studies employing a ^{14}C-assay similar to that described by Gold (1964) have revealed a similar degree of versatility in the symbionts *P. convolutae*, *A. chattonii*, and *A. klebsii*. Of particular interest is the utilization of urea, ammonia, uric acid, glycerophosphate, and adenylic, guanylic, and cytadylic acids.

The ability of symbionts to utilize organic substrates heterotrophically could be of great significance, and is currently being examined in cultures with the screening technique of Parsons and

Strickland (1962). Early results suggest that *G. microadriaticum*, *P. convolutae*, *A. klebsii*, and *A. chattonii* are all capable of varying degrees of heterotrophy. Compounds commonly assimilated include acetate, succinate, pyruvate, lactate, and citrate (Taylor, unpublished data). Comparisons between *P. convolutae* and the experimental symbionts employed by Provasoli, Yamasu, and Manton (1968) are of interest, since the unnatural symbionts show varying degrees of heterotrophy which are reflected in their success as symbionts. *Tetraselmis verrucosa* and *Platymonas* sp. (Plymouth 315) are unable to incorporate labeled organic compounds and are by far the worst symbionts. In contrast, *Prasinocladus marinus* (Plymouth 308) shows the greatest potential heterotrophic ability. It is of interest that *P. marinus* was a better symbiont (in terms of host growth and reproduction) than *P. convolutae* (Provasoli, Yamasu, and Manton 1968). These results suggest that algal heterotrophy may be an important symbiont character, functioning in the association to provide efficient nutrient cycling. In situations where light is limiting, heterotrophy could spare photosynthesis, thereby allowing hosts such as hermatypic corals to exist at considerable depths.

The quality and quantity of the photosynthate produced by the symbionts and the amount that is translocated to the host play an important role in the survival of the association. Much of what is known of this aspect of algal symbiosis has been obtained through the in vitro methods of Muscatine and his colleagues (Muscatine 1967; Smith, Muscatine, and Lewis 1969; Trench 1971a, b, c; Muscatine, Pool, and Cernichiari 1972), who have studied a variety of associations involving *G. microadriaticum*. Principal translocated products include glycerol, alinine, glucose, fumarate, succinate, and glycollate. This does vary depending upon the host species. In the majority of cases the amount of material translocated can approach 40 to 50% of the total photosynthetically incorporated carbon. Studies of *G. microadriaticum* in culture show a similar range of compounds being excreted, but the amount is generally less than 5% of the total carbon fixed. The range of compounds excreted does not vary among strains of the symbiont isolated from different hosts. Attempts to raise the level of excretion in culture have been unsuccessful (Raylor 1971a). The elucidation of a postulated "host factor" (Muscatine 1967) may eventually overcome this problem.

The principal compounds excreted by *P. convolutae* include glucose, fructose, mannitol, and lactic acid (Taylor 1971a and unpublished data). These have been shown to be utilized by alga-free hosts. The rate of excretion by *P. convolutae* is dependent on membrane permeability in the alga, and can be artificially increased by lowering the pH of the medium (Taylor 1971a). A similar mechanism is suspected to operate in the intact association.

DISCUSSION

As more algal symbionts are brought into culture, the most striking feature to emerge is their apparent uniformity regardless of host
type. As seen here, this is particularly true of *G. microadriaticum*.
It serves to emphasize the biological virtuosity of this organism,
not its apparent cultural blandness. Important questions can be
answered by examining the adaptive potential of this and other symbiotic algae in culture. For example, it should be possible to
experiment with factors which control the rate of excretion of photosynthate, to study the effects of light quality on the character of
translocated materials and to determine the alga's capacity to synthesize pharmacologically active compounds such as the diterpene
lactones of the gorgonians. Systems of experimentally induced artificial symbioses employing cultured cells should aid in understanding host preferences and the adaptive abilities of various symbionts.
In short, the biology of algal-invertebrate symbiosis can be effectively studied through a careful manipulation of intact associations
and their separated components.

ACKNOWLEDGMENTS

The views expressed here are the direct result of studies supported by the Browne Fund of the Royal Society (Naples, Italy, 1968;
Discovery Bay, Jamaica, 1972), and NSF-NATO postdoctoral fellowship
and NSF grants GB19790, GB19790A. I am particularly grateful to the
Directors of the Marine Biological Association of the United Kingdom,
the Naples Zoological Station and the U.W.I.-S.U.N.Y. Marine Laboratory, Discovery Bay, for their generous provision of laboratory space
and facilities.

Figures 5 and 6 are reproduced courtesy of the Council of the
Marine Biological Association of the United Kingdom.

This paper is No. 0000 from the Rosenstiel School of Marine and
Atmospheric Science, University of Miami, Miami, Florida.

LITERATURE CITED

Apelt, G. 1969. Die Symbiose zwischen dem acoelen Turbellar *Convoluta convoluta* Diatomeen der Gattung *Licmophora*. Mar. Biol.
3:165-87.

Ax, P., and Apelt, G. 1965. Die "Zooxanthellen von *Convoluta con-
voluta* (Turbellaria:Acoela) entstehen ans Diatomeen. Erster
Nacheis einer Endosymbiose zwischen Tieren und Kiesel-algen.
Naturwissenschaften 52:444-46.

Brandt, K. 1883. Über die morphologische und physiologische Bedeu-
tung des Chlorophylls bei Tieren. Mit. Zool. Stn. Neapel 4:
191-302.

Caullery, M. 1952. <u>Parasitism</u> <u>and</u> <u>symbiosis</u>. London: Sidgwick and Jackson.

Chatton, E. 1923. Les Péridiniens parasites des Radiolaires. <u>C</u>. <u>R</u>. <u>Acad</u>. <u>Sci</u>., Paris 177:1246–49.

Christensen, T. 1965. Systematisk Botanik,nr. 2, Alger. <u>Botanik</u> 2:1–180.

Dodge, J. D. 1968. The fine structure of chloroplasts and pyrenoids in some marine dinoflagellates. <u>J</u>. <u>Cell</u> <u>Sci</u>. 3:41–48.

_____, and Crawford, R. M. 1970. A survey of thecal fine structure in the Dinophyceae. <u>Bot</u>. <u>J</u>. <u>Linn</u>. <u>Soc</u>. 63:53–67.

Doyle, W. L., and Doyle, M. M. 1940. The structure of zooxanthellae. <u>Pap</u>. <u>Tortugas</u> <u>Lab</u>. 32:127–42.

Droop, M. R. 1963. Algae and invertebrates in symbiosis. In <u>Symbiotic association</u>, ed. P. S. Nutman and B. Mosse, pp. 171–99. Cambridge: Cambridge University Press.

Freudenthal, H. D. 1962. *Symbiodinium* gen. nov. and *Symbiodinium microadriaticum* sp. nov., a zooxanthella: taxonomy, life cycle, and morphology. <u>J</u>. <u>Protozool</u>. 9:45–52.

_____, Lee, J. J., and Kossoy, V. 1964. Cytochemical studies of the zooxanthella from the planktonic foraminifera *Globigerinoides ruber*. <u>J</u>. <u>Protozool</u>. 11:12.

Fritsch, F. E. 1935. <u>The</u> <u>structure</u> <u>and</u> <u>reproduction</u> <u>of</u> <u>the</u> <u>algae</u>, Vol. 1. Cambridge: Cambridge University Press.

_____. 1952. Algae in association with heterotrophic and holozoic organisms. <u>Proc</u>. <u>Roy</u>. <u>Soc</u>. (Lond.) B139:85–192.

Gamble, F. W., and Keeble, F. 1904. The bionomics of *Convoluta roscoffensis* with special reference to its green cells. <u>Quart</u>. <u>J</u>. <u>Microsc</u>. <u>Sci</u>. 47:363.

Geddes, P. 1880. Sur la chlorophylle animale et la fonction des Planaires vertes. <u>Arch</u>. <u>Zool</u>. <u>Exp</u>. <u>et</u> <u>Gen</u>. 8:51–62.

Gold, K. 1964. A microbiological assay for vitamin B-12 in seawater using radiocarbon. <u>Limnol</u>. <u>Oceanog</u>. 9:343–47.

Goreau, T. F., Goreau, N. I., and Yonge, C. M. 1971. Reef corals: autotrophs or heterotrophs? <u>Biol</u>. <u>Bull</u>. 141:247–60.

Hovasse, R. 1923. *Endodinium chattonii* (nov. gen., nov. sp.). Parasite des Vélleles. Un type exceptionnel de variation du nombre des chromosomes. <u>Bull</u>. <u>Biol</u>. <u>Fr</u>. <u>Belg</u>. 58:34–38.

Kawaguti, S. 1944. On the physiology of reef corals. VII. Zooxanthellae of the reef corals is *Gymnodinium* sp., dinoflagellate; its culture <u>in</u> <u>vitro</u>. <u>Palao</u> <u>Trop</u>. <u>Biol</u>. <u>Stn</u>. <u>Stud</u>. 2:675–79.

Keeble, F. 1910. <u>Plant</u> <u>animals</u>. Cambridge: Cambridge University Press.

_____, and Gamble, F. W. 1907. The origin and nature of the green cells of *Convoluta roscoffensis*. <u>Quart</u>. <u>J</u>. <u>Microsc</u>. <u>Sci</u>. 51:167–219.

Kevin, M. J., Hall, W. T., McLaughlin, J. J., and Zahl, P. A. 1969. *Symbiodinium microadriaticum* Freudenthal, a revised taxonomic description, ultrastructure. <u>J</u>. <u>Phycol</u>. 5:341–50.

Kinsey, R. A. 1970. <u>The</u> <u>ecology</u> <u>of</u> <u>the</u> <u>gorgonians</u> (Cnidaria, Octocorallia) <u>of</u> <u>Discovery</u> <u>Bay</u>, <u>Jamaica</u>. Ph. D. dissertation, Yale University.

Klebs, G. 1884. Ein kleiner Beitrag zur Kenntniss der Peridinien. Bot. Zeit. 10:46–67.

Lebour, M. V. 1930. The Planktonic diatoms of northern seas. London: Roy. Soc.

Lee, J. J., and Zucker, W. 1969. Algal flagellate symbiosis in the Foraminifer Archaias. J. Protozool. 16:71–80.

_____, Freudenthal, H. D., Kossoy, V., and Bé, A. 1965. Cytological observations on the two planktonic foraminifera, Globigerina bulloides d'Orbigny, 1826, and Globigerinoides ruber (d'Orbigny, 1839) Cushman, 1927. J. Protozool. 12:531–42.

McLaughlin, J. J. A., and Zahl, P. A. 1957. Studies in marine biology. II. In vitro culture of zooxanthellae. Proc. Soc. Exptl. Biol. Med. 95:115–20.

_____. 1959. Axenic zooxanthellae from various invertebrate hosts. Ann. N. Y. Acad. Sci. 77:55–72.

_____. 1962a. Axenic cultivation of the dinoflagellate symbiont from the coral Cladocora. Arch. Mikrobiol. 42:40–41.

_____. 1962b. Endozoic algae. In The physiology and biochemistry of algae, ed. R. A. Lewin, pp. 823–26. New York: Academic Press.

_____. 1966. Endozoic algae. In Symbiosis, ed. S. M. Henry, Vol. 1, pp. 257–97. New York: Academic Press.

Muscatine, L. 1967. Glycerol excretion by symbiotic algae from corals and Tridacna and its control by the host. Science 156:516–19.

_____, Pool, R. R., and Cernichiari, E. 1972. Some factors influencing selective release of soluble organic material by zooxanthellae from reef corals. Mar. Biol. 13:298–308.

Norris, R. 1967. Micro-algae in enrichment cultures from Puerto Penasco, Sonora, Mexico. Bull. S. Calif. Acad. Sci. 66:233–50.

Nozawa, K., Taylor, D. L., and Provasoli, L. 1972. Respiration and photosynthesis in Convoluta roscoffensis Graff, infected with various symbionts. Biol. Bull. 143:420–30.

Oschman, J. L. 1966. Development of the symbiosis of Convoluta roscoffensis Graff and Platymonas sp. J. Phycol. 2:105–11.

Parke, M., and Manton, I. 1967. The specific identity of the algal symbiont in Convoluta roscoffensis. J. Mar. Biol. Ass. U. K. 47:445–64.

Parson, T. R., and Strickland, J. D. H. 1962. On the production of particulate organic carbon by heterotrophic processes in seawater. Deepsea Res. 8:211–22.

Pringsheim, E. G. 1955. Die "gelben Zellen" der Koralle Cladocora. Pubbl. Staz. Zool., Napoli 27:5–9.

Provasoli, L., McLaughlin, J. J. A., and Droop, M. R. 1957. The development of artificial media for marine algae. Arch. Mikrobiol. 25:392–428.

_____, Yamasu, T., and Manton, I. 1968. Experiments on the resynthesis of symbiosis in Convoluta roscoffensis with different flagellate cultures. J. Mar. Biol. Ass. U. K. 48:465–79.

Sará, M. 1971. Ultrastructural aspects of the symbiosis between two species of the genus *Aphanocapsa* (Cyanophyceae) and *Ircinia variabilis* (Demospongiae). Mar. Biol. 11:214-21.

_____, and Liaci, L. 1964. Associazione fra la Cianoficea *Aphanocapsa feldmanni* e alcune Demospongie marine. Boll. Zool. 31: 55-65.

Smith, D. C., Muscatine, L., and Lewis, D. H. 1969. Carbohydrate movement from autotrophs to heterotrophs in parasitic and mutualistic symbiosis. Biol. Rev. 44:17-90.

Taylor, D. L. 1968. In situ. studies on the cytochemistry and ultrastructure of a symbiotic marine dinoflagellate. J. Mar. Biol. Ass. U. K. 48:349-66.

_____. 1969a. Identity of zooxanthellae isolated from some Pacific Tridacnidae. J. Phycol. 5:336-40.

_____. 1969b. Studies on the taxonomy of some symbiotic marine dinoflagellates. In Progess in protozoology, ed. A. A. Strelkov, K. M. Sukhanova, and Raikov, I. B., p. 373. Leningrad: Third Int. Congr. Protozool.

_____. 1969c. The ultrastructure of *Endodinium chattonii* Hovasse a symbiotic marine dinoflagellate. In Progress in protozoology, ed. A. A. Strelkov, K. M. Sukhanova, and Raikov, I. B., p. 76. Leningrad: Third Int. Congr. Protozool.

_____. 1971a. Patterns of carbon translocation in algal-invertebrate symbiosis, pp.590-97. Sapporo, Japan: Proceedings VII International Seaweed Symposium. U. Tokyo Press.

_____. 1971b. On the symbiosis between *Amphidinium klebsii* (Dinophyceae) and *Amphiscolops langerhansi* (Turbellaria:Acoela). J. Mar. Biol. Ass. U. K. 51:301-14.

_____. 1971c. Ultrastructure of the "zooxanthella" *Endodinium chattonii* in situ. J. Mar. Biol. Ass. U. K. 51:227-34.

_____. 1971d. Taxonomy of some common *Amphidinium* species. Br. Phycol. J. 6:129-33.

_____. 1973. The cellular interactions of algal-invertebrate symbiosis. Advances in Marine Biology 11 (In press).

Trench, R. K. 1971a. The physiology and biochemistry of zooxanthellae symbiotic with marine coelenterates. I. Assimilation of photosynthetic products of zooxanthellae by two marine coelenterates. Proc. Roy Soc. (Lond.) B177:225-35.

_____. 1971b. The physiology and biochemistry of zooxanthellae symbiotic with marine coelenterates. II. Liberation of fixed ^{14}C by zooxanthellae in vitro. Proc. Roy. Soc. (Lond.) B177: 237-50.

_____. 1971c. The physiology and biochemistry of zooxanthellae symbiotic with marine coelenterates. III. The effect of homogenates of host tissues on the excretion of photosynthetic products in vitro by zooxanthellae from two marine coelenterates. Proc. Roy. Soc. (Lond.) B177:251-64.

Winter, F. W. 1907. Zur Kenntniss der Thalamorphoren. Arch. Protistenk 10:1-113.

Zahl, P. A., and McLaughlin, J. J. A. 1957. Isolation and cultivation of zooxanthellae. Nature 180:199-200.

———. 1959. Studies in marine biology. IV. On the role of algal cells in the tissues of invertebrates. J. Protozool. 6:344-52.

SUMMARY

The papers in this symposium represent a wide range of studies on symbiotic relationships in marine organisms. The issues raised fall into two general areas of discussion: host-symbiont interactions and evolutionary trends.

Vandermeulen and Muscatine set the stage for discussion of the complexities of symbiont-host relationships with their presentation on the symbiotic dinoflagellates inhabiting reef-building corals. While algae are known to influence many facets of reef coral biology, such as productivity, nutrient recycling, and rate of deposition of coral skeleton, the mechanisms are incompletely understood. Vandermeulen and Muscatine experimentally tested the theories advanced to explain the role of the symbiotic algae in accelerating calcification in reef corals. Although they were able to clarify some of the relationships between coral calcification and the symbiotic algae, the authors point out that more experiments are needed before conceptual models of coral calcification can be developed further.

Another plant-animal association that has received attention in recent years involves the chloroplasts found in opisthobranch molluscs. Although it has been recognized for many years that species of opisthobranchs contain chlorophyll-like pigments in their bodies, only recently have these pigments been identified as long-lived algal chloroplasts. Greene pointed out evidence of photosynthetic activity in these structures, and suggested that they are, therefore, functional. Many unanswered questions, however, remain. It is not known, for example, whether or not the chloroplasts can synthesize their own chlorophyll pigments, or even if they can reproduce themselves. Neither is it known if the chloroplasts are symbiotic only in animal cells or if they are also symbiotic in plants.

A highly interesting problem in symbiotic relationships is the manner in which host-symbiont contact is made. In a series of studies on the symbiotic shrimp, *Beateaus macginitieae* and its urchin host *Strongylocentrotus franciscanus*, Ache found that symbiont contact could be made either visually or chemically, but efficiency in host location was greatest when both cues were present. In other studies on lower invertebrate symbionts, chemical orientation proved to be the chief controlling mechanism in host location. However, Ache postulates that among the higher invertebrates the stimuli of other modalities complements, or perhaps even replaces, the role of chemoreception in host location.

Once host-symbiont contact has been established, many of the problems faced by symbionts are the same as those of the free-living animals. To answer the question of the spatial distribution patterns of a symbiont on its host, Dimock investigated the territoriality of the polychaete *Arctone* on its sea cucumber host *Stichopus parvimensis* and the limpet *Megathum crenulata*. Small symbionts were random in distribution; larger symbionts, however, were found as isolated individuals usually uniformly distributed near the oral surface. Dimock concluded that the intraspecific aggression allowed each worm to maintain an isolated distributional pattern, thus reducing competition for a common resource, in this case probably the ample supply of food found on the oral surface of the host.

Endosymtionts are also faced with the problem of population regulation. At least 8 species of trematode larval parasites utilize the digestive gland and gonads of the mud-flat snail *Nassarius obsoletus*, and at times two or more species are found in the same host snail. DeCoursey and Vernberg asked how two species that appear to be competing for the same resources can coexist. Their investigations indicated a number of possible interactions among the competing larval helminth species. In two of the species there seemed to be coexistence without harmful interaction, although it was not known if this was a long-term relationship. In other pairs there was a preferential selection of different microhabitats within the gonad-digestive areas or displacement to other tissues. Certain species of larval trematodes are never found together in the same host, possibly because of cannibalism or biochemical interactions.

Although population growth and regulation in symbionts are governed by many of the same factors as those in free-living animals, there is the additional factor of host resistance. Thus, the ability of the host to recognize "self" and "non-self" is critical to the establishment of an endosymbiont. One aspect of this concept is phagocytosis, as illustrated by the relationship between marine molluscs and their endosymbionts. The granulocytes of at least some species of marine molluscs are generally extremely efficient in recognizing "self" from "non-self", thus smaller foreign materials are readily phagocytized. Cheng et al suggest that this is the reason the Microsporidea, which are widely distributed in invertebrates are rarely found in molluscs. Those that do occur are hyperparasites, and thus are immune to phagocytic action of the host hemolymph cells.

The evolutionary aspects of symbiotic relationships have long puzzled biologists. In an attempt to shed light on trends in the development of the anemone-crab/mollusc association, Ross examined four types of data: systematics, zoogeography, behavior, and ecology. Very few taxonomic groups are involved in these symbioses. The crustacean hosts are not limited to any particular taxonomic family, while the anemone symbionts, on the other hand, are limited to the family Harmathiidae. Geographically the crab-anemone associations are limited to the warmer seas and shallow waters. In contrast, gastropod-anemone partnerships are found in boreal and cold temperature conditions. Certain general types of behavior have been observed in the establishment of these associations: activity of the anemone itself in transferring to hermit crab shells and living molluscs, active detachment of anemones by crabs using legs and claws to relax and ease off the symbiont, and the relaxation of the anemones in response to the stimulation by a prospective crab host. Ecologically the anemone may be a passenger only, or it may protect the crab by either forming a cloak or actively protecting its host.

A number of interesting questions were raised about this relationship. At what stage in the life cycle does the species pair come together? How did the association evolve? Were the first associations anemone-molluscs or anemone-crabs?

Continuing in the quest for evolutionary trends, Jennings examined symbiotic relationships among the turbellarians. All types of associations can be found in this group of animals, ranging from temporary associations to obligate parasitism. The plasticity of the turbellarian stock enables them to exploit a wide range of habitats, and Jennings interpreted this plasticity as a preadaptation to a symbiotic life style. Although the symbiotic turbellarians generally retain the same patterns of structure and physiology as the free-living forms, several adaptations foreshadow the extreme modifications in trematodes and cestodes. Thus, the different symbioses found in the turbellaria demonstrate a possible model for cestode and trematode evolution. The symbiosis probably arose through association of the symbiont with an appropriate host, at first by chance or in search of a temporary shelter, then later leading to ecto- and entocommensalism.

The origin of the digenetic trematodes, their taxonomy, the evolution of life cycles, and the phylogeny of the Digenea were covered in depth by Cable. The origin and phylogeny of the aspidobothreans was also considered. Further life history and comparative morphology studies are needed to clarify the systematic position of many species. Cable also suggested that determination of amount and complexity of DNA would offer insight into the origin and evolution of trematodes. Such studies hold promise of making precise genome comparisons, elucidating the origin and subsequent evolutionary lines of the digenetic trematodes.

During the evolution of the digenetic trematodes, they have become adapted to remarkably diverse habitats. Vernberg and Vernberg pinpointed some of the physiologically adaptive responses of the trematodes that have enabled them to exploit these habitats. In some aspects each stage apparently has evolved more or less independently of the next. Each stage must be able to utilize the energy resources provided and at the same time be able to withstand the environmental fluctuations associated with its host. Other physiological responses, such as response to temperature, are the same throughout the life cycle of the species.

One of the most controversial parasitic flatworms in terms of taxonomy and phylogeny is the genus *Gyrocotyle*. Found only in chimaeroid fish, each species is restricted to a single host species. Usually only a single pair of worms survive from a large initial infection. Free-living larvae have been reported but little is known of their biology. Even more puzzling are the post-larval stages which have been found in the tissues of adult worms. Whether the animals use an intermediate host or whether they penetrate the fish host directly has not been established. The worms have been generally classified as cestoids, principally because they lack a digestive tract. The gyrocotylid epidermis is also similar to that of the cestodes, but this could be a result of endoparasitism and loss of a gut.

The full spectrum of symbiotic relationships can be found in branching coral, *Pocillopora damicornis*. Of the four groups of associates found with *P. damicornis* - agile shrimp, sedentary shrimp, crabs, and fish - no one species predominates. The symbiont species show varying degrees of adaptation to the host. Some show no structural modifications, and appear to be able to live equally well on or off the coral. One group has evolved feeding habits that makes it possible for them to feed without leaving the coral, although they do not feed on coral tissue itself. Other species have become obligate symbionts with many structural adaptations as well as a possible dependency on host mucus. Finally, the fish *Paragobiodon echinocephalus* relies on coral tissue as its major food source, and hence can be regarded as parasitic.

Why are some species of organisms better adapted for symbiosis than others? Taylor indicated that among the algae, successful symbionts have either structural or nutritional adaptations. The green algae *Platymonas convolutae*, for example, has a naked protoplast that allows nutrients and waste materials to be readily transported between host and algae. Algal hetrotrophy is also an important factor in successful symbionts. It provides efficient nutritional cycling, and where light is limiting, allows the animal to live in the absence of photosynthesis.

These papers have touched on many facets of biology and have updated current knowledge of symbiotic relationships. Although there are obviously many areas of research that will further clarify some of the unanswered questions pinpointed at this symposium, three general research fields hold promise of being particularly fruitful: behavior and sensory physiology, physiological ecology, and biochemistry. Development of research in these three areas is in part a natural consequence of technological advances that permit development of quantitative data and increased interest in the ecological significance of symbiosis.

INDEX